Biofuels: Types and Production

Biofuels: Types and Production

Michael Silva

Larsen & Keller
www.larsen-keller.com

Biofuels: Types and Production
Michael Silva
ISBN: 978-1-64172-651-1 (Hardback)

🖿 Larsen & Keller

Published by Larsen and Keller Education,
5 Penn Plaza,
19th Floor,
New York, NY 10001, USA

Cataloging-in-Publication Data

Biofuels : types and production / Michael Silva.
    p. cm.
Includes bibliographical references and index.
ISBN 978-1-64172-651-1
1. Biomass energy. 2. Microbial fuel cells. 3. Refuse as fuel.
4. Waste products as fuel. I. Silva, Michael.
TP339 .B56 2022
662.88--dc23

For more information regarding Larsen and Keller Education and its products, please visit the publisher's website www.larsen-keller.com

TABLE OF CONTENTS

Chapter 5 Production Processes and Technologies 190

Permissions

Index

This book is a culmination of my many years of practice in this field. I attribute the success of this book to my support group. I would like to thank my parents who have showered me with unconditional love and support and my peers and professors for their constant guidance.

Biofuel is a fuel that is produced in a relatively small time interval from biomass by using modern techniques. It can be derived from plants known as energy crops as well as from the biological wastes produced by agriculture, households and industries. Some of the commonly used biofuels are biodiesel, bioethanol, biogas, syngas, methanol, butanol, bioethers and green diesel. Biodiesel is produced through the process of transesterification by using fats or oils. Bioethanol is produced by the process of fermentation using the carbohydrates which are present in sugar or starch crops. Biofuels are classified into various generations, namely, first generation, second generation, third generation and fourth generation. The topics covered in this extensive book deal with the core subjects of biofuel. It elucidates new techniques and their applications in a multidisciplinary approach. This book will serve as a reference to a broad spectrum of readers.

The details of chapters are provided below for a progressive learning:

Chapter – Introduction

The fuel that is produced from biomass by various modern methods rather than biological processes is known as biofuel. It is majorly used in transportation, heat and power generation. This is an introductory chapter which will introduce briefly all the significant aspects of biofuels including its uses, advantages and disadvantages.

Chapter – Generations of Biofuels

Biofuels have been divided into four generations on the basis of their sources, namely, first, second, third and fourth. These biofuels are derived from food crops, biomass, algae and electrofuels respectively. The topics elaborated in this chapter will help in gaining a better perspective about these generations of biofuels.

Chapter – Types of Biofuels

Some of the common types of biofuels are biogas, syngas, algae fuel , biodiesel, vegetable oil fuel, biohydrogen, bioethers, alcohol fuel, aviation biofuel and biogasoline. This chapter has been carefully written to provide an easy understanding of these types of biofuels.

Chapter – Feedstocks for Biofuels

Feedstock is the unprocessed material that is used to supply a manufacturing process. Some of the common feedstocks are Camelina, Jatropha Curcas, Lignocellulosic Biomass, Switchgrass and Miscanthus Giganteus. The diverse use of these energy crops as feedstocks for biofuels have been thoroughly discussed in this chapter.

Chapter – Production Processes and Technologies

Some of the most common processes and techniques for biofuel production include dark fermentation, photofermentation, vegetable oil refining and gasification. The chapter closely examines these key concepts, technologies and processes related to biofuel production to provide an extensive understanding of the subject.

Michael Silva

Introduction

- **Uses of Biofuel**

- **Energy Content of Biofuel**

- **Advantages of Biofuel**

- **Disadvantages of Biofuel**

- **Biomass**

The fuel that is produced from biomass by various modern methods rather than biological processes is known as biofuel. It is majorly used in transportation, heat and power generation. This is an introductory chapter which will introduce briefly all the significant aspects of biofuels including its uses, advantages and disadvantages.

Biofuel is any fuel that is derived from biomass—that is, plant or algae material or animal waste. Since such feedstock material can be replenished readily, biofuel is considered to be a source of renewable energy, unlike fossil fuels such as petroleum, coal, and natural gas. Biofuel is commonly advocated as a cost-effective and environmentally benign alternative to petroleum and other fossil fuels, particularly within the context of rising petroleum prices and increased concern over the contributions made by fossil fuels to global warming. Many critics express concerns about the scope of the expansion of certain biofuels because of the economic and environmental costs associated with the refining process and the potential removal of vast areas of arable land from food production.

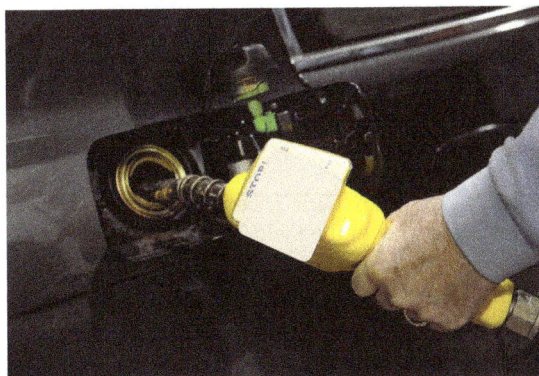

Ethanol gas fuel pump delivering the E85 mixture to an automobile in Washington state, U.S.

Types of Biofuels

Some long-exploited biofuels, such as wood, can be used directly as a raw material that is burned to produce heat. The heat, in turn, can be used to run generators in a power plant to produce electricity. A number of existing power facilities burn grass, wood, or other kinds of biomass.

An ethanol production plant in South Dakota, U.S.

Liquid biofuels are of particular interest because of the vast infrastructure already in place to use them, especially for transportation. The liquid biofuel in greatest production is ethanol (ethyl alcohol), which is made by fermenting starch or sugar. Brazil and the United States are among the leading producers of ethanol. In the United States ethanol biofuel is made primarily from corn (maize) grain, and it is typically blended with gasoline to produce "gasohol," a fuel that is 10 percent ethanol. In Brazil, ethanol biofuel is made primarily from sugarcane, and it is commonly used as a 100-percent-ethanol fuel or in gasoline blends containing 85 percent ethanol. Unlike the "first-generation" ethanol biofuel produced from food crops, "second-generation" cellulosic ethanol is derived from low-value biomass that possesses a high cellulose content, including wood chips, crop residues, and municipal waste. Cellulosic ethanol is commonly made from sugarcane bagasse, a waste product from sugar processing, or from various grasses that can be cultivated on low-quality land. Given that the conversion rate is lower than with first-generation biofuels, cellulosic ethanol is dominantly used as a gasoline additive.

Algal Biofuel: Research technician, Nick Sweeney, inoculates algae being grown in a tent reactor in the algal lab in the Field Test Laboratory Building (FTLB).

The second most common liquid biofuel is biodiesel, which is made primarily from oily plants (such as the soybean or oil palm) and to a lesser extent from other oily sources (such as waste cooking fat from restaurant deep-frying). Biodiesel, which has found greatest acceptance in Europe, is used in diesel engines and usually blended with petroleum diesel fuel in various percentages.

The use of algae and cyanobacteria as a source of "third-generation" biodiesel holds promise but has been difficult to develop economically. Some algal species contain up to 40 percent lipids by weight, which can be converted into biodiesel or synthetic petroleum. Some estimates state that algae and cyanobacteria could yield between 10 and 100 times more fuel per unit area than second-generation biofuels.

Other biofuels include methane gas and biogas—which can be derived from the decomposition of biomass in the absence of oxygen—and methanol, butanol, and dimethyl ether—which are in development.

Economic and Environmental Considerations

In evaluating the economic benefits of biofuels, the energy required to produce them has to be taken into account. For example, the process of growing corn to produce ethanol consumes fossil fuels in farming equipment, in fertilizer manufacturing, in corn transportation, and in ethanol distillation. In this respect, ethanol made from corn represents a relatively small energy gain; the energy gain from sugarcane is greater and that from cellulosic ethanol or algae biodiesel could be even greater.

Biofuels also supply environmental benefits but, depending on how they are manufactured, can also have serious environmental drawbacks. As a renewable energy source, plant-based biofuels in principle make little net contribution to global warming and climate change; the carbon dioxide (a major greenhouse gas) that enters the air during combustion will have been removed from the air earlier as growing plants engage in photosynthesis. Such a material is said to be "carbon neutral." In practice, however, the industrial production of agricultural biofuels can result in additional emissions of greenhouse gases that may offset the benefits of using a renewable fuel. These emissions include carbon dioxide from the burning of fossil fuels during the production process and nitrous oxide from soil that has been treated with nitrogen fertilizer. In this regard, cellulosic biomass is considered to be more beneficial.

Land use is also a major factor in evaluating the benefits of biofuels. The use of regular feedstock, such as corn and soybeans, as a primary component of first-generation biofuels sparked the "food versus fuel" debate. In diverting arable land and feedstock from the human food chain, biofuel production can affect the economics of food price and availability. In addition, energy crops grown for biofuel can compete for the world's natural habitats. For example, emphasis on ethanol derived from corn is shifting grasslands and brushlands to corn monocultures, and emphasis on biodiesel is bringing down ancient tropical forests to make way for oil palm plantations. Loss of natural habitat can change the hydrology, increase erosion, and generally reduce biodiversity of wildlife areas. The clearing of land can also result in the sudden release of a large amount of carbon dioxide as the plant matter that it contains is burned or allowed to decay.

Some of the disadvantages of biofuels apply mainly to low-diversity biofuel sources—corn, soybeans, sugarcane, oil palms—which are traditional agricultural crops. One alternative involves the use of highly diverse mixtures of species, with the North American tallgrass prairie as a specific example. Converting degraded agricultural land that is out of production to such high-diversity biofuel sources could increase wildlife area, reduce erosion, cleanse waterborne pollutants, store carbon dioxide from the air as carbon compounds in the soil, and ultimately restore fertility to degraded lands. Such biofuels could be burned directly to generate electricity or converted to liquid fuels as technologies develop.

The proper way to grow biofuels to serve all needs simultaneously will continue to be a matter of much experimentation and debate, but the fast growth in biofuel production will likely continue. In the United States the Energy Independence and Security Act of 2007 mandated the use of 136 billion litres (36 billion gallons) of biofuels annually by 2022, more than a sixfold increase over 2006 production levels. The legislation also requires, with certain stipulations, that 79 billion litres (21 billion gallons) of the total amount be biofuels other than corn-derived ethanol, and it continued certain government subsidies and tax incentives for biofuel production.

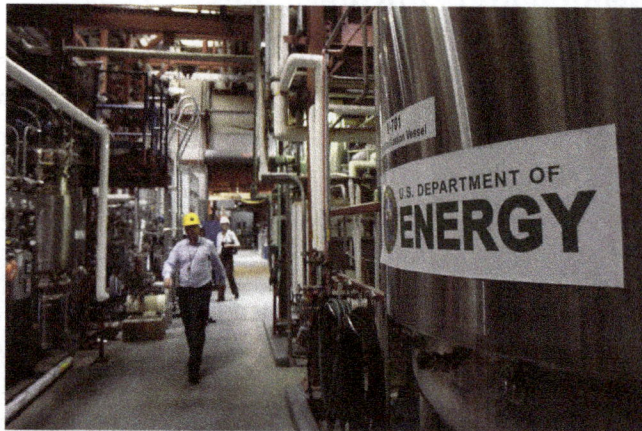

Biofuels Testing Centre: Workers at the biofuels testing centre at the National Renewable Energy Laboratory (NREL).

One distinctive promise of biofuels is that, in combination with an emerging technology called carbon capture and storage, the process of producing and using biofuels may be capable of perpetually removing carbon dioxide from the atmosphere. Under this vision, biofuel crops would remove carbon dioxide from the air as they grow, and energy facilities would capture the carbon dioxide given off as biofuels are burned to generate power. Captured carbon dioxide could be sequestered (stored) in long-term repositories such as geologic formations beneath the land, in sediments of the deep ocean, or conceivably as solids such as carbonates.

Uses of Biofuel

When using the work fuel, the tendency is the think of the material in question as a liquid for use in automobiles, trucks, and possibly planes. The truth is, biofuels are looked at as a means of replacing ALL of human energy needs from home heating to vehicle fuel to electricity generation. The basic concept is that if we use as much product as we grown, then our net impact on the environment should be negligible if not zero.

Transportation

Nearly 30% of all energy consumed in the United States is used in transportation. To put this into perspective, residential and commercial uses combined only account for 10%. That means that humans in industrial nations use, on average, three times more energy to get around than they use to cook their food and heat their homes. This number does not include electricity generation, which accounts for 40% of all energy used.

Globally, transportation accounts for 25% of energy demand and nearly 62% of oil consumed. Most of this energy , two-thirds in fact, is burned to operate vehicles with the rest going to maintenance, manufacturing, infrastructure, and raw material harvesting. If we delve further into the numbers, we find that upwards of 70% of energy consumption in this segment is used to move people around and that most of this is used in private cars, the least efficient means of transportation. Only 12% of the energy burned by a car goes to moving it and only about 2% is actually used to move the occupants. The rest of the energy is lost to friction, heat, inefficient combustion, and moving about ever more heavy vehicles.

Estimates are that we have hit peak oil or if we have not, it is very near. We won't actually know that we have peaked until we start down the slope toward the bottom again, but most experts agree that we are quite close. So, oil is running short and when this is combined with the tremendous environmental impact of petroleum recovery, refining, and eventual combustion, the drive for an alternative is clear.

The problem with many alternatives, like wind, solar, etc. is that they simply aren't practical. Transporting enough stored electricity derived from these mechanisms to make an average journey is very difficult. Many experts believe that practical breakthroughs in these technologies are decades away at best. So, the challenge is to find a fuel that can replace the practical qualities of oil (like being easy to drive around), but which does not pollute the same way.

The solution, at least for now, appears to be algal-based biofuels, which are still years if not decades away from commercialization. The idea is simple. Algae have lipid and lipid can be converted to a number of fuels including diesel, ethanol, butanol, and methanol. Because algae absorb CO_2 to make lipid, the net impact on the environment should be very small. Additionally, biofuels are biodegradable, so if they do spill, less harm is done compared to when fossil fuels spill. What is the hold-up you ask? At this point in time, developing fuel from algae requires huge investments of water and fertilizer because the algae must be killed in order to harvest the lipid and then a new stock is grown back up again. The energy needed to grow algae from a seed stock to "harvest-ready" is orders of magnitude larger than the energy obtained from harvesting them. In other words, more energy is put into the system than is taken out, so it leads to a net loss. Until the input of energy is lower than what the system produces (excluding energy from the sun), the system will not be viable.

Power Generation

The generation of electricity is the single largest use of fuel in the world. In 2008, the world produced about 20,261 TWh of electricity. About 41% of that energy came from coal, another 21% came from natural gas, and the rest was covered by hydro, nuclear, and oil at 16%, 13%, and 5% respectively. Of the fuel burned, only 39% went into producing energy and rest was lost as heat. Only 3% of the heat was then used for co-generation. Of the 20,261 TWh produced, 16,430 TWh were delivered to consumers and the rest was used by the plants themselves.

It is clear that a great deal of energy goes into producing electricity, which isn't surprising given that everything humans do in the industrialized world, from running water to surfing the internet, requires electricity. Most estimates suggest that about 40% of all GHG emissions come from the production of electricity, with transportation coming in a very close second. Coal, in particular, is highly problematic for its production of sulfur dioxide, which produces acid rain. Interestingly, nuclear power is the least damaging in terms of pollutants produced, generating less carbon than any form of power generation other than hydro and including solar (PV panel production uses large amounts of water).

So, if humans are not going to switch to nuclear power, then a cleaner, more renewable form of energy is needed. Biofuels may provide at least a partial answer. Co-generation plants often use methane derived from landfills and there is vigorous interest in the use of syngas in many agricultural areas. Like any biofuel, the balance of the equation lies in carbon generation. For syngas made from the agricultural waste, the net impact is lower than if the waste were allowed to decompose on its own. This is because natural decomposition in oxygen-rich environments produces nitrogen dioxide, with is over 300 times more potent of a greenhouse gas than carbon dioxide, as well as methane, which is over 20 times more potent. The same benefits exist for methane harvested from landfills.

these applications are not enough to meet our energy needs and so the conversion of crops grown specifically for energy is where most of the research and development is occurring at this stage. Algae and other plants that grown in harsh conditions and thus do not threaten the food supply are actively under investigation for potential sources of biofuel. At this point, only about 13% of all electricity in the United States is made from renewable sources (excluding hydro), but very little of this is biofuel. Most of the electricity from biofuels is produced as a byproduct of fuel production for transportation. The United Kingdom is the largest market for biofuel-to-electricity generation, generating enough power for 350,000 households from landfill gas alone.

Heat

The major use of natural gas from fossil fuels is heat, though a good deal of it also goes to energy. In the United States, a boom in hydraulic fracturing (called Fracking) has led to a huge surge in the production of natural gas from shale (a fossil fuel) and to the prediction that this will soon become the predominant form of energy, perhaps as soon as 2040. Of course, natural gas need not come from fossilized plant material, it can also be produced from recently grown plant material.

Of course, the majority of biofuel used in heating is solid. Wood is both an aesthetic and a practical method of heating and may homes use wood burning stoves as supplements to other heating systems like natural gas or electricity. Renewed interest in solid biofuels, in part a response to rising energy prices, as led to a surge in innovation in the industry with research focusing on improved efficiency, reduced emissions, and enhanced convenience. Wood gasification boilers can reach efficiencies as high as 91%.

To put the cost of biofuel into perspective, 1,000 BTUs of energy from wood cost about $1.20. Natural gas, on the other hand, cost about $2.60 per 1,000 BTUs. Wood pellets cost around $2.16 per 1,000 BTUs, making them less expensive than natural gas as well. The table below shows the cost of various fuels and provides a not on efficiency.

Fuel	Cost per 1,000 BTUs in U.S. Dollars	Energy Efficiency
Heating Oil	$3.17	78%
Natural Gas	$2.60	80%
Wood	$1.20	70-91%
Electricity	$3.22	100% (in home only) 40% (from plant)
Propane	$3.13	78%
Wood Pellets	$2.16	70%
Coal	$1.10	70%

Energy Content of Biofuel

The Energy content of biofuel is a description of the chemical energy contained in a given biofuel, measured per unit mass of that fuel, as specific energy, or per unit of volume of the fuel, as energy density. A biofuel is a fuel, produced from living organisms. Biofuels include bioethanol, an alcohol made by fermentation—often used as a gasoline additive, and biodiesel, which is usually used as a diesel additive. Specific energy is energy per unit mass, which is used to describe the energy content of a fuel, expressed in SI units as joule per kilogram (J/kg) or equivalent units. Energy density is the amount of energy stored in a fuel per unit volume, expressed in SI units as joule per litre (J/L) or equivalent units.

Energy and CO_2 Output of Common Biofuels

The table below includes entries for popular substances already used for their energy, or being discussed for such use.

The second column shows specific energy, the energy content in megajoules per unit of mass in kilograms, useful in understanding the energy that can be extracted from the fuel.

The third column in the table lists energy density, the energy content per liter of volume, which is useful for understanding the space needed for storing the fuel.

The final two columns deal with the carbon footprint of the fuel. The fourth column contains the proportion of CO_2 released when the fuel is converted for energy, with respect to its starting mass, and the fifth column lists the energy produced per kilogram of CO_2 produced. As a guideline, a higher number in this column is better for the environment. But these numbers do not account for other green house gases released during burning, production, storage, or shipping. For example, methane may have hidden environmental costs that are not reflected in the table.

Fuel Type	Specific energy (MJ/kg)	Energy Density (MJ/L)	CO_2 Gas made from Fuel Used (kg/kg)[nb 1]	Energy per CO_2 (MJ/kg)
Solid Fuels				
Bagasse (Cane Stalks)	9.6		$\sim$+40%$(C_6H_{10}O_5)_n$+15%$(C_{26}H_{42}O_{21})_n$+15%$(C_9H_{10}O_2)_n$1.30	7.41
Chaff (Seed Casings)	14.6		–	
Animal Dung/Manure	10- 15		–	
Dried plants $(C_6H_{10}O_5)_n$	10 – 16	1.6 - 16.64	IF50%$(C_6H_{10}O_5)_n$+25%$(C_{26}H_{42}O_{21})_n$+25%$(C_{10}H_{12}O_3)_n$1.84	5.44-8.70
Wood fuel $(C_6H_{10}O_5)_n$	16 – 21	2.56 - 21.84	IF45%$(C_6H_{10}O_5)_n$+25%$(C_{26}H_{42}O_{21})_n$+30%$(C_{10}H_{12}O_3)_n$1.88	8.51-11.17
Charcoal	30		85-98% Carbon+VOC+Ash 3.63	8.27
Liquid Fuels				
Pyrolysis oil	17.5	21.35	(Assumption Of Fuel: Carbon Content = 23% w/w) 0.84	20.77
Methanol $(CH_3\text{-}OH)$	19.9 – 22.7	15.9	1.37	14.49-16.53
Ethanol $(CH_3\text{-}CH_2\text{-}OH)$	23.4 – 26.8	18.4 - 21.2	1.91	12.25-14.03
Ecalene™	28.4	22.7	75%C_2H_6O+9%C_3H_8O+7%$C_4H_{10}O$+5%$C_5H_{12}O$+4%Hx 2.03	14.02
Butanol$(CH_3\text{-}(CH_2)_3\text{-}OH)$	36	29.2	2.37	15.16
Fat	37.656	31.68	–	
Biodiesel	37.8	33.3 – 35.7	~2.85	~13.26
Sunflower oi $(C_{18}H_{32}O_2)$	39.49	33.18	(12%$(C_{16}H_{32}O_2)$+16%$(C_{18}H_{34}O_2)$+71%(LA)+1%(ALA))2.81	14.04
Castor oil $(C_{18}H_{34}O_3)$	39.5	33.21	(1%PA+1%SA+89.5%ROA+3%OA+4.2%LA+0.3%A-LA)2.67	14.80
Olive oil $(C_{18}H_{34}O_2)$	39.25 - 39.82	33 - 33.48	(15%$(C_{16}H_{32}O_2)$+75%$(C_{18}H_{34}O_2)$+9%(LA)+1%(ALA))2.80	14.03
Gaseous Fuels				
Methane (CH_4)	55 – 55.7	(Liquefied) 23.0 – 23.3	(Methane leak exerts 23 × greenhouse effect of CO_2) 2.74	20.05-20.30
Hydrogen (H_2)	120 – 142	(Liquefied) 8.5 – 10.	(Hydrogen leak slightly catalyzes ozone depletion) 0.0	
Fossil Fuels (comparison)				
Coal	29.3 – 33.5	39.85 - 74.43	(Not Counting:CO, NO_x, Sulfates & Particulates) ~3.59	~8.16-9.33
Crude Oil	41.868	28 – 31.4	(Not Counting:CO,NO_x,Sulfates & Particulates) ~3.4	~12.31
Gasoline	45 – 48.3	32 – 34.8	(Not Counting:CO,NO_x,Sulfates & Particulates) ~3.30	~13.64-14.64
Diesel	48.1	40.3	(Not Counting:CO,NO_x,Sulfates & Particulates) ~3.4	~14.15
Natural Gas	38 – 50	(Liquefied) 25.5 – 28.	(Ethane, Propane & Butane N/C:CO,NO_x & Sulfates) ~3.00	~12.67-16.67
Ethane $(CH_3\text{-}CH_3)$	51.9	(Liquefied) ~24.0	2.93	17.71
Nuclear fuels (comparison)				
Uranium-235 (^{235}U)	77,000,000	(Pure)1,470,700,000	[Greater for lower ore conc.(Mining, Refinin , Moving)] 0.0	~55 - ~90
Nuclear fusion ($^2H\text{-}^3H$)	300,000,000	(Liquefied)53,414,377.	(Sea-Bed Hydrogen-Isotope Mining-Method Dependent) 0.0	
Fuel Cell Energy Storage (comparison)				
Direct-Methanol	4.5466	3.6	~1.37	~3.31
Proton-Exchange (R&D)	up to 5.68	up to 4.5	(IFF Fuel is recycled) 0.0	
Sodium Hydride (R&D)	up to 11.13	up to 10.24	(Bladder for Sodium Oxide Recycling) 0.0	
Battery Energy Storage (comparison)				

Lead-acid battery	0.108	~0.1	(200-600 Deep-Cycle Tolerance) 0.0	
Nickel-iron battery	0.0487 - 0.1127	0.0658 - 0.1772	(<40y Life)(2k-3k Cycle Tolerance IF no Memory effect) 0.0	
Nickel-cadmium battery	0.162 - 0.288	~0.24	(1k-1.5k Cycle Tolerance IF no Memory effect) 0.0	
Nickel metal hydride	0.22 - 0.324	0.36	(300-500 Cycle Tolerance IF no Memory effect) 0.0	
Super iron battery	0.33	(1.5 * NiMH) 0.54	(~300 Deep-Cycle Tolerance) 0.0	
Zinc-air battery	0.396 - 0.72	0.5924 - 0.8442	(Recyclable by Smelting & Remixing, not Recharging) 0.0	
Lithium ion battery	0.54 - 0.72	0.9 - 1.9	(3-5 y Life) (500-1k Deep-Cycle Tolerance) 0.0	
Lithium-Ion-Polymer	0.65 - 0.87	(1.2 * Li-Ion)1.08 - 2.28	(3-5 y Life) (300-500 Deep-Cycle Tolerance) 0.0	
Lithium iron phosphate battery		–		–
DURACELL Zinc-Air	1.0584 - 1.5912	5.148 - 6.3216	(1-3 y Shelf-life) (Recyclable not Rechargeable) 0.0	
Aluminium battery	1.8 - 4.788	7.56	(10-30 y Life) (3k+ Deep-Cycle Tolerance) 0.0	
PolyPlusBC Li-Aircell	3.6 - 32.4	3.6 - 17.64	(May be Rechargeable)(Might leak sulfates) 0.0	

1. While all CO_2 gas output ratios are calculated to within a less than 1% margin of error(assuming total oxidation of the carbon content of fuel), ratios preceded by a Tilde (~) indicate a margin of error of up to (but no greater than) 9%. Ratios listed do not include emissions from fuel plant cultivation/Mining, purification/refining and transportation. Fuel availability is typically 74–84.3% NET from source Energy Balance.

2. While Uranium-235 (^{235}U) fission produces no CO_2 gas directly, the indirect fossil fuel burning processes of Mining, Milling, Refining, Moving & Radioactive waste disposal, etc. of intermediate to low-grade uranium ore concentrations produces some amount of carbon dioxide. Studies vary as to how much carbon dioxide is emitted. The United Nations Intergovernmental Panel on Climate Change reports that nuclear produces approximately 40 g of CO_2 per kilowatt hour (11 g/MJ, equivalent to 90 MJ/kg CO_2e). A meta-analysis of a number of studies of nuclear CO_2 lifecycle emissions by academic Benjamin K. Sovacool finds nuclear on average produces 66 g of CO_2 per kilowatt hour (18.3 g/MJ, equivalent to 55 MJ/kg CO_2e). One Australian professor claims that nuclear power produces the equivalent CO_2 gas emissions per MJ of net-output-energy of a Natural Gas fired power station.

Table: Yields of common crops associated with biofuels production.

Crop	Oil (kg/ha)	Oil (L/ha)	Oil (lb/acre)	Oil (US gal/acre)	Oil per seeds (kg/100 kg)	Melting Range (°C) Oil / Fat	Melting Range (°C) Methyl Ester	Melting Range (°C) Ethyl Ester	Iodine number	Cetane number
Groundnut					(Kernel)42					
Copra					62					
Tallow						35 - 42	16	12	40 - 60	75
Lard						32 - 36	14	10	60 - 70	65

Crop	Oil (kg/ha)	Oil (L/ha)	Oil (lb/acre)	Oil (US gal/acre)	Oil per seeds (kg/100 kg)	Oil / Fat	Methyl Ester	Ethyl Ester	Iodine number	Cetane number
Corn (maize)	145	172	129	18		-5	-10	-12	115 - 124	53
Cashew nut	148	176	132	19						
Oats	183	217	163	23						
Lupine	195	232	175	25						
Kenaf	230	273	205	29						
Calendula	256	305	229	33						
Cotton	273	325	244	35	(Seed)13	-1 - 0	-5	-8	100 - 115	55
Hemp	305	363	272	39						
Soybean	375	446	335	48	14	-16 - -12	-10	-12	125 - 140	53
Coffee	386	459	345	49						
Linseed (flax)	402	478	359	51		-24			178	
Hazelnuts	405	482	362	51						
Euphorbia	440	524	393	56						
Pumpkin seed	449	534	401	57						
Coriander	450	536	402	57						
Mustard seed	481	572	430	61	35					
Camelina	490	583	438	62						
Sesame	585	696	522	74	50					
Safflower	655	779	585	83						
Rice	696	828	622	88						
Tung oil tree	790	940	705	100		-2.5			168	
Sunflower	800	952	714	102	32	-18 - -17	-12	-14	125 - 135	52
Cocoa (cacao)	863	1,026	771	110						
Peanuts	890	1,059	795	113		3			93	
Opium poppy	978	1,163	873	124						
Rapeseed	1,000	1,190	893	127	37	-10 - 5	-10 - 0	-12 - -2	97 - 115	55 - 58
Olives	1,019	1,212	910	129		-12 - -6	-6	-8	77 - 94	60
Castor beans	1,188	1,413	1,061	151	(Seed)50	-18			85	
Pecan nuts	1,505	1,791	1,344	191						
Jojoba	1,528	1,818	1,365	194						
Jatropha	1,590	1,892	1,420	202						
Macadamia nuts	1,887	2,246	1,685	240						
Brazil nuts	2,010	2,392	1,795	255						
Avocado	2,217	2,638	1,980	282						
Coconut	2,260	2,689	2,018	287		20 - 25	-9	-6	8 - 10	70
Chinese Tallow[nc 2]		4,700		500						
Oil palm	5,000	5,950	4,465	635	20-(Kernal)36	20 - 40	-8 - 21	-8 - 18	12 - 95	65 - 85
Algae		95,000		10,000						

1. Typical oil extraction from 100 kg of oil seeds.

2. Chinese Tallow (Sapium sebiferum, or Tradica Sebifera) is also known as the "Popcorn Tree".

Advantages of Biofuel

Availability of Biofuels

Unlike fossil fuels, biofuels are a renewable energy source. Because they are derived from crops that can be harvested annually, or in the case of algae monthly, biofuels are theoretically unlimited. Despite this surface appearance of unlimited availability, biofuels do have restrictions. Restrictions are treated in more depth in disadvantages of biofuels, but a brief consideration reveals that the threat to the food supply is the major limiting factor to the quantity of biofuel feedstock can be grown.

This limitation also means that certain feedstocks are out of the running for replacing fossil fuels. Crops like corn and soybeans do not produce enough energy per acre of crop to meet current fuel needs, which are only expected to increase, without seriously threatening the food supply. For this reason, higher energy density crops like algae and Jatropha are being considered.

An abstraction of availability is delivery infrastructure. After all, a fuel that it easily produced but not easily transported (like electricity from solar panels in the Sahara) is still limited in its availability. Biofuels are similar in many ways to fossil fuels. They are liquid at standard temperature and pressure, have reasonably high energy densities, and can be distributed with only minor modifications to existing infrastructure.

Speaking of modifications, biofuels have the advantage that they can be burned in standard internal combustion engines with only minor modifications to the rubber in fuel lines and gaskets. This is in stark contrast to fuels like hydrogen or electricity, which requires complete redesign of everything from the engine to the transmission.

So, in terms of availability, biofuels have a big advantage as they are at the top of the list of alternatives and, as supplies slowly dwindle, will also top fossil fuels. Availability may be the driving force in adoption of alternatives energies, making biofuels the next logical choice while other alternatives are still under development. In fact, biofuels are already showing up in full fuel engines, in countries like Brazil, and as additives to standard fossil fuels in almost every nation. The transition is likely to be subtle but slow as more and more fossil fuel is replaced with biofuel. The U.S. military, for instance, plans to replace 50% of its fossil-based jet fuel with biofuel alternatives by 2016.

Environmental Impact

This category is tricky because biofuels are very similar to hydrocarbons and have some of the same emissions problems that standard fossil fuels have. They can, however, be more environmentally friendly if care is taken in how they are produced and distributed. It is also the case that biofuels have an impact on the environment other than emissions, so we must consider several different subcategories under this heading.

Spills and Surface Contamination

Biofuels are not 100% safe but they are much safer than fossil fuels. If you were to spill a large quantity of biofuel into a concentrated area, it would likely kill living organisms and contaminate surround soil or water. However, the scale of the impact would be orders of magnitude smaller than with fossil fuels.

First off, biofuels are biological molecules and this means they are biodegradable. Bacteria and other organisms that live naturally in the soil and water are able to use biofuel molecules as energy sources and break them down into harmless byproducts. This means that even though concentrated biofuel spills can kill things like plants and smaller animals, they will not persist in the environment and cause damage or make an area uninhabitable for long periods of time.

Sulphur and Atmospheric Contamination

One of the major problems to arise from burning fossil fuels, especially coal, is acid rain that comes from the high sulphur content of these fuels. Biofuels can be produced in ways that completely eliminate sulphur and thus can eliminate this component of acid rain.

On the other hand, biofuels tend to contain high levels of nitrogen, which can form compounds that also lead to acid rain and atmospheric contamination. On the whole, the net impact on acid rain production is usually negative, meaning biofuels can reduce acid rain. Importantly, biofuels can be carefully produced to ensure that contamination is as low as possible, giving them an edge over fossil fuels because it is easier to avoid contamination in the production phase than it is to remove contaminants during refining.

Greenhouse Gas (GHG) Emissions and Global Warming

This is the area in which the most care must be taken in how biofuels are produced. If biofuels are produced in the "correct" way, they can greatly reduce greenhouse gas emissions. If produced incorrectly, they can increase emissions.

First, plants use carbon dioxide, the major greenhouse gas of concern, to grow and produce food. So, plants are able to reduce the amount of carbon dioxide in the atmosphere and thus decrease global warming. Biofuels, when grown from plants, can thus offset their CO_2 admissions because they take up the gas during growth that is produced when the fuel is burned. The idea is that if there is a one-to-one relationship, then the gas produced is the same as the gas taken in and there is no net impact on global warming. The problem is that achieving the one-to-one ratio may be impossible.

For starters, energy has to be invested into growing the crop itself. This energy comes in the form of planting seeds, tilling and preparing the ground, and importing water and nutrients. As it turns out, you cannot get something for nothing and so many crops require more energy input than they give out in the end. In other words, if you take into account the GHG emissions that occur just to grow the crop and add that to the greenhouse gas emissions from burning the crop, there is more

CO_2 produced than taken up and global warming worsens. As of yet, there is no good solution to this problem. Many companies are looking to invest energy in the form of sunlight so that there is no GHG emitted during the production phase. There is still a net energy input, but no greenhouse gas is produced. This seems to be most feasible with algae.

The other problem to consider is land use. If land is cleared to grow a biofuel, then the plant life that existed there is eliminated. This problem is considered in more detail in the disadvantages of biofuel, but the main point is that carbon is produced to clear that land and the benefits of the plants on the land are lost. By some estimates and depending on the type of plant life removed, the impact could be a carbon debt that can take as long as 500 years to pay back. Again, the solution to this problem may be algae.

If the above technical impediments can be overcome, then the net impact of biofuels on the environment can be limited. In such a scenario, the greenhouse gas emissions and impact on global warming will be far lower with biofuels than with fossil fuels. The feasibility of achieving this advantage remains to be seen.

Energy Independence

This advantage is obvious and has no immediate drawbacks. If a country has the land resources to grow biofuel feedstock, then it can produce its own energy. This ends any dependence on fossil fuel resources, which are geographically limited to only a few places in the world. Given the amount of conflict that occurs over fuel supplies and prices, energy independence should have a net positive effect.

Despite this utopian ideal, the reality of biofuel energy independence is not so clear cut. First, not every country has the resources needed to grow biofuels. Many countries do not have the land area, access to water, or ability to produce fertilizer for crops and thus would still need to rely on others for their fuel to some degree.

As a second point, the shift in power could have a highly disruptive effect. First, national economies around the world depend on oil revenue to survive. Many Middle Eastern countries have a vested interest in ensuring that oil remains important and profitable given that as much as 90% of government revenue in these places comes from oil exports. To compound this problem, most of these countries would go from net energy exporters to net energy importers, further damaging their economies and forcing them to completely shift their industrial and commercial focuses.

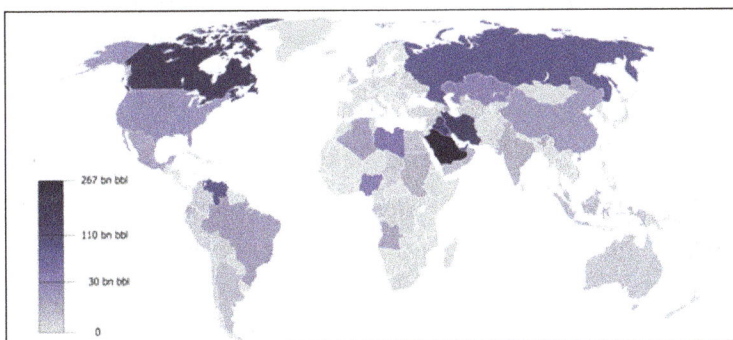

Major Oil Producing Regions.

Finally, other countries with vast land resources and access to good growing conditions stand to become hew hotbeds of conflict. It is unlikely that the fight over energy will cease given that the location of suppliers will change and not everyone will be able to meet their energy needs from biofuels.

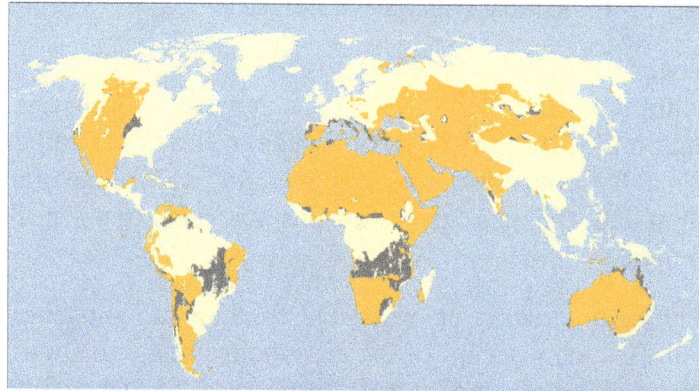

Dry Regions of World.

In the map above, dry regions are shown in grey. Red regions indicate areas that are not considered desert, but are at high risk of becoming desert. Blue regions are too cold to grow crops.

Disadvantages of Biofuel

Though biofuels have a number of advantages over fossil fuels, their integration into the fuel supply chain has to be done with great care to ensure that their potential disadvantages are, if not eliminated, at least minimized.

At the current time, more "radical" alternative fuel technologies such as solar and wind have one major problem - portability. It is very difficult to transport large quantities of electricity without using heavy, relatively inefficient batteries. Biofuels, on the other hand, are fairly easy to transport, have decent energy densities, and can be used with only minor modifications to existing technology and infrastructure. With that in mind, it seems that biofuels will likely act as a stop gap or filler that helps us as we transition from fossil fuels to other technologies that aren't quite ready. We attempt to outline the disadvantages that should be considered when a biofuel is being implemented, not to discourage their use, but rather to encourage responsible use of this technology, which is likely to become our predominant source of energy as we transition over to cleaner alternatives in the more distant future. With that in mind, here are the disadvantages to be aware of.

Regional Suitability

Despite pushes for Jatropha, Camelina, and algae, it is more likely that biofuel feedstock will be grown on a regional basis. This is important for a number of reasons, chief among them being the fact that some crops just grow better in some locations and not so well in other. Things that will need to be taken into consideration are:

- Water use - The less water a crop uses, the better as water is a very limited resource. This is particularly important in places that are more arid.

- Invasiveness - A crop that kills native plants and which is difficult to control is not a good choice as it may threaten biodiversity and severely damage the surrounding ecosystem.

- Fertilizer - Nutrients are needed for plants to grow. Some plants are more frugal with scarce resources than others.

- Limitations - Some places just aren't going to be able to grow biofuel crops. Alaska, for instance, really isn't suited to the rapid growth needed to produce crops year after year for fuel supplies. Regions like this will have to import fuels, so energy independence still will not be possible for every location.

Food Security

Biofuel feedstock has to be grown and there is only so much suitable land in the world for growing plants. Very little, for instance, is going to grow in the Sahara Desert. The problem with growing crops for fuel is that they take up land that could be used for growing food. In a world with a population of around 7 billion and that is already short on food, there will necessarily be a tradeoff between food crop and biofuel feedstock. Every effort is made to grow feedstock that uses "none agricultural land." This means crops like corn and soybeans are out of the running.

Now, the impact on the food supply may not be quite as great as initially estimated for the simple reason that land classified as "agricultural," isn't necessarily being actively planted. This has occurred as advances in crop technology have led to the production of larger harvests on less land. Estimates in 2008 suggested that the following regions could grow biofuel on abandoned agricultural land.

Region	Biofuel Production (millions of tonnes/yr)
Africa	88 - 245
Asia	139 - 293
Australia/Indonesia	95 - 321
Europe	144 - 364
North America	211 - 697
South America	154 - 480

The problem is as difficult to ascertain as it is to solve. Some estimates suggest that if biofuel production doubles from its 2006 production, then by the year 2020 there will be an additional 90 million people at risk of hunger on top of those already at risk. This should indicate that even though farmland sits unused, that does not mean that there is abundant food across the planet. The truth may be closer to the fact that unused farmland is not profitable for food production, even though tens of millions of people are at risk of hunger right now. The vast majority of the increase in "at risk for hunger" individuals, if biofuel production increases, will occur in Eastern Asia, but there will still be 20 million people in "developed" countries who will be additionally put at risk. The major driver of this increase in hunger risk comes from increases in food prices that will result from the fact that agricultural land can "earn more" if it is planted with biofuels. That means farmers will demand higher prices for food to offset what they lose by not planting biofuel feedstock.

Land use Changes

This one is little more difficult to understand, but you can think of it as akin to deforestation. If the land used to grow a biofuel feedstock has to be cleared of native vegetation, then ecological damage is done in three ways.

The first way damage is caused is by destroying local habitat. This necessarily destroys animal dwellings, microcosms (micro ecosystems), and reduces the overall health of a region's natural resources. The loss of plant life also means that the world loses valuable CO_2 scrubbers. Even though the land is being replanted, a native forest is almost always better at removing CO_2 from the atmosphere than a biofuel feedstock, in part because the CO_2 remains trapped and is never released by burning as with fuel stock.

The second way that damage is done is in the carbon debt created. Energy is needed to deforest an area and prepare it for farming as well as to plant the crop. All of this leads to the production of greenhouse gases and puts the region at a net positive GHG production before a single biofuel is even produced. Then, energy must be invested into harvesting the replanting the crop the next time around. Estimates have shown that deforesting native land can actually produce a carbon debt that can take up to 500 years to repay. So, using native land for biofuels, even if no food is grown on it, puts us in the hole ecologically.

Finally, as if the two problems above where not enough, changing land to agricultural status almost always means fertilizers are going to be used. It only makes sense to use fertilizer if we want to get the most yield per area. The problem is runoff and other agricultural pollution, which rivals that of urban pollution in its impact on the local environment. Thus, creating more farmland is likely to damage waterways and require us to invest energy into treatment plants and other mitigation strategies. The net result is an even larger carbon debt.

As one can see, land use changes for biofuel production are a very, very bad way to go and should be avoided at all costs. The best solution is to use existing land, but that puts food supplies at risk. So the problem is very difficult to solve. Some people have proposed using algae, which grows in very inhospitable regions and has limited impact on land use. The problem with algae, however, is water use.

Monoculture, Genetic Engineering and Biodiversity

It is easier to grow a large quantity of a single crop if it is all very uniform. This is referred to as monoculture and examples can be seen in the corn, soybean, and potato farming sectors. Potatoes in the U.S., for instance, are almost always Russet potatoes because that is what is in demand by large consumers like fast food chains. The problems with growing a single crop over large tracts of land are several-fold.

First, growing only one crop changes the environment in terms of the food available to pests. This is an evolutionary pressure that can lead to a number of problems. For instance, if a crop of potatoes is eaten by a certain pest that can only migrate a few hundred feet and the potato fields are separated by corn fields, then an outbreak in one potato field is not a problem because it won't spread. Without the corn fields, however, the pest is free to destroy an entire crop.

Now, for the second problem, we could treat the pests mentioned above with pesticide, but it is inevitable that a few of those pests will be resistant. After all, out of the hundreds of thousands or even millions of insects, bacteria, and fungi that can inhabit a single field of crops, at least a few are likely (by chance alone) to be resistant to the chemicals we use to kill them. After all, we can't spray too much as it would be damaging to human health. So the result is a pest that is resistant to pesticide. Now it is free to eat all the crop it wants and we are powerless to stop it.

The next problem comes when we turn to genetic engineering. We decide to modify the crop so that it is resistant to the pest without the need for pesticides. The problem is, the same things happens because it is likely at least a few pests aren't affected by the modification or a new pest comes along and we are left, after a few years, with the same problem as when we began.

The point is, there are limits to how much of a single crop can be grown and there is little that technology can do about that. The key to healthy crops worldwide is biodiversity, which simply means having lots of different types of plants and animals around. That way, if the Russet potatoes suffer blight, we still have Yukon Gold or Red Thumb potatoes to turn to. This is especially important when dealing with food crops. Just ask the Irish and they'll tell you how much havoc a pest can wreak in a food supply built on a monoculture.

Global Warming

This probably goes without saying and won't be belabored here, but burning biofuels, which are most hydrogen and carbon, produce carbon dioxide, which contributes to global warming. So, even though biofuels may be able to help ease our energy needs, they won't solve all of our problems.

Now, it may be true that biofuels produce LESS GHG emissions than fossil fuels, but that can only serve to slow global warming and not to stop or reverse it. Thus, biofuels can only be substitutes for the short term as we invest in other technologies. The key to implementing them is to mitigate environmental impact by being mindful of the disadvantages.

Biomass

Biomass is plant or animal material used for energy production, heat production, or in various industrial processes as raw material for a range of products. It can be purposely grown energy crops (e.g. miscanthus, switchgrass), wood or forest residues, waste from food crops (wheat straw, bagasse), horticulture (yard waste), food processing (corn cobs), animal farming (manure, rich in nitrogen and phosphorus), or human waste from sewage plants.

Burning plant-derived biomass releases CO_2, but it has still been classified as a renewable energy source in the EU and UN legal frameworks because photosynthesis cycles the CO_2 back into new crops. In some cases, this recycling of CO_2 from plants to atmosphere and back into plants can even be CO_2 negative, as a relatively large portion of the CO_2 is moved to the soil during each cycle.

Cofiring with biomass has increased in coal power plants, because it makes it possible to release less CO_2 without the cost associated with building new infrastructure. Co-firing is not without issues however, often an upgrade of the biomass is beneficiary. Upgrading to higher grade fuels can be achieved by different methods, broadly classified as thermal, chemical, or biochemical.

Biomass Feedstocks

Biomass plant.

Wood waste outside biomass power plant.

Bagasse is the remaining waste after sugar canes have been crushed to extract their juice.

Miscanthus x giganteus energy crop.

Historically, humans have harnessed biomass-derived energy since the time when people began burning wood fuel. Even in 2019, biomass is the only source of fuel for domestic use in many developing countries. All biomass is biologically-produced matter based in carbon, hydrogen and oxygen. The estimated biomass production in the world is approximately 100 billion metric tons of carbon per year, about half in the ocean and half on land.

Wood and residues from wood, for instance spruce, birch, eucalyptus, willow, oil palm, remains the largest biomass energy source today. It is used directly as a fuel or processed into pellet fuel or other forms of fuels. Biomass also includes plant or animal matter that can be converted into fuel, fibers or industrial chemicals. There are numerous types of plants, including corn, switchgrass, miscanthus, hemp, sorghum, sugarcane, and bamboo. The main waste energy feedstocks are wood waste, agricultural waste, municipal solid waste, manufacturing waste, and landfill gas. Sewage sludge is another source of biomass. There is ongoing research involving algae or algae-derived biomass. Other biomass feedstocks are enzymes or bacteria from various sources, grown in cell cultures or hydroponics.

Based on the source of biomass, biofuels are classified broadly into two major categories:

- First-generation biofuels are derived from food sources, such as sugarcane and corn starch. Sugars present in this biomass are fermented to produce bioethanol, an alcohol fuel which serve as an additive to gasoline, or in a fuel cell to produce electricity.

- Second-generation biofuels utilize non-food-based biomass sources such as perennial energy crops (low input crops), and agricultural/municipal waste. There is huge potential for second generation biofuels but the resources are currently under-utilized.

Biomass Conversion

Thermal Conversions

Straw bales.

Thermal conversion processes use heat as the dominant mechanism to upgrade biomass into a better and more practical fuel. The basic alternatives are torrefaction, pyrolysis, and gasification, these are separated principally by the extent to which the chemical reactions involved are allowed to proceed (mainly controlled by the availability of oxygen and conversion temperature).

There are other less common, more experimental or proprietary thermal processes that may offer benefits, such as hydrothermal upgrading. Some have been developed for use on high moisture content biomass, including aqueous slurries, and allow them to be converted into more convenient forms.

Chemical Conversion

A range of chemical processes may be used to convert biomass into other forms, such as to produce a fuel that is more practical to store, transport and use, or to exploit some property of the process itself. Many of these processes are based in large part on similar coal-based processes, such as the Fischer-Tropsch synthesis. Biomass can be converted into multiple commodity chemicals.

Biochemical Conversion

As biomass is a natural material, many highly efficient biochemical processes have developed in nature to break down the molecules of which biomass is composed, and many of these biochemical conversion processes can be harnessed. In most cases, microorganisms are used to perform the conversion process: anaerobic digestion, fermentation, and composting.

Glycoside hydrolases are the enzymes involved in the degradation of the major fraction of biomass, such as polysaccharides present in starch and lignocellulose. Thermostable variants are gaining increasing roles as catalysts in biorefining applications, since recalcitrant biomass often needs thermal treatment for more efficient degradation.

Electrochemical Conversion

Biomass can be directly converted to electrical energy via electrochemical (electrocatalytic) oxidation of the material. This can be performed directly in a direct carbon fuel cell, direct liquid fuel cells such as direct ethanol fuel cell, a direct methanol fuel cell, a direct formic acid fuel cell, a L-ascorbic Acid Fuel Cell (vitamin C fuel cell), and a microbial fuel cell. The fuel can also be consumed indirectly via a fuel cell system containing a reformer which converts the biomass into a mixture of CO and H_2 before it is consumed in the fuel cell.

Environmental Impact

On combustion, the carbon from biomass is released into the atmosphere as carbon dioxide (CO_2). After a period of time ranging from a few months to decades, the CO_2 produced from combustion is absorbed from the atmosphere by plants or trees. However, the carbon storage capacity of forests may be reduced overall if destructive forestry techniques are employed.

All biomass crops sequester carbon. For example, soil organic carbon has been observed to be greater below switchgrass crops than under cultivated cropland, especially at depths below 30 cm (12 in). For Miscanthus x giganteus, McCalmont et al. found accumulation rates ranging from 0.42 to 3.8 tonnes per hectare per year, with a mean accumulation rate of 1.84 tonne (0.74 tonnes per acre per year), or 20% of total harvested carbon per year. The grass sequesters carbon in its continually increasing root biomass, toghether with carbon input from fallen leaves. Typically, perennial crops sequester more carbon than annual crops because the root buildup is allowed to continue undisturbed over many years. Also, perennial crops avoid the yearly tillage procedures (plowing, digging) associated with growing annual crops. Tilling induces soil aeration, which accelerates the soil carbon decomposition rate, by stimulating soil microbe populations. Also, tilling makes it easier for the oxygen (O) atoms in the atmosphere to attach to carbon (C) atoms in the soil, producing CO_2.

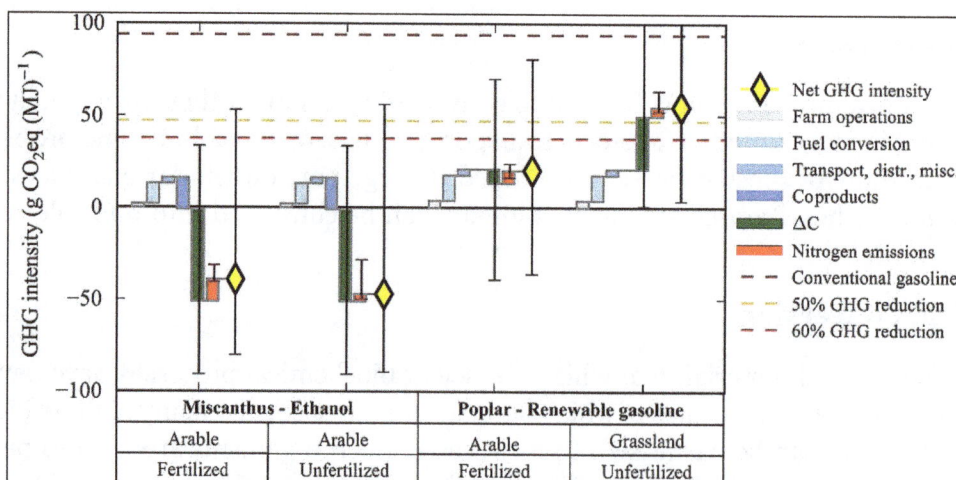

In the above figure is shown GHG/CO_2/carbon negativity for Miscanthus x giganteus production pathways. Relationship between above-ground yield (diagonal lines), soil organic carbon (X axis), and soil's potential for successful/unsuccessful carbon sequestration (Y axis). Basically, the higher the yield, the more land is usable as a GHG mitigation tool.

The simple proposal that biomass is carbon-neutral put forward in the early 1990s has been superseded by the more nuanced proposal that for a particular bioenergy project to be carbon neutral, the total carbon sequestered by a bioenergy crop's root system must compensate for all the emissions from the related, aboveground bioenergy project. This includes any emissions caused by direct or indirect land use change. Many first generation bioenergy projects are not carbon neutral given these demands. Some have even higher total GHG emissions than some fossil based alternatives. Transport fuels might be worse than solid fuels in this regard.

Some are carbon neutral or even negative, though, especially perennial crops. The amount of carbon sequestrated and the amount of GHG (greenhouse gases) emitted will determine if the total GHG life cycle cost of a bio-energy project is positive, neutral or negative. Specifically, a GHG/carbon negative life cycle is possible if the total below-ground carbon accumulation more than compensates for the above-ground total life-cycle GHG emissions. Whitaker et al. estimates that for Miscanthus x giganteus, carbon neutrality and even negativity is within reach. Basically, the yield and related carbon sequestration is so high that it more than compensates for both farm operations emissions, fuel conversion emissions and transport emissions. The graphic on the right displays two CO_2 negative Miscanthus x giganteus production pathways, represented in gram CO_2-equivalents per megajoule. The yellow diamonds represent mean values.

One should note that successful sequestration is dependent on planting sites, as the best soils for sequestration are those that are currently low in carbon. The varied results displayed in the graph highlights this fact. Milner et al. argues that for the UK, successful sequestration is expected for arable land over most of England and Wales, with unsuccessful sequestration expected in parts of Scotland, due to already carbon rich soils (existing woodland). Also, for Scotland, the relatively lower yields in this colder climate makes CO_2 negativity harder to achieve. Soils already rich in carbon includes peatland and mature forest. Grassland can also be carbon rich, however Milner et al. further argues that the most successful carbon sequestration in the UK takes place below improved grasslands. The bottom graphic displays the estimated yield necessary to achieve CO_2 negativity for different levels of existing soil carbon saturation.

Forest-based biomass projects has received criticism for ineffective GHG mitigation from a number of environmental organizations, including Greenpeace and the Natural Resources Defense Council. Environmental groups also argue that it might take decades for the carbon released by burning biomass to be recaptured by new trees. Biomass burning produces air pollution in the form of carbon monoxide, volatile organic compounds, particulates and other pollutants. In 2009 a Swedish study of the giant brown haze that periodically covers large areas in South Asia determined that two thirds of it had been principally produced by residential cooking and agricultural burning, and one third by fossil-fuel burning. The use of wood biomass as an industrial fuel has been shown to produce fewer particulates and other pollutants than the burning seen in wildfires or open field fires.

References

- Biofuel, technology: britannica.com, Retrieved 13 August, 2019
- Kenneth e. Heselton (2004), "boiler operator's handbook". Fairmont press, 405 pages. Isbn 0881734357
- Uses-of-biofuels: biofuel.org.uk, Retrieved 21 July, 2019

- Intergovernmental panel on climate change (2007). "4.3.2 nuclear energy". Ipcc fourth assessment report: climate change 2007, working group iii mitigation of climate change. Retrieved 2011-02-07

- Advantages-of-biofuels: biofuel.org.uk, Retrieved 29 July, 2019

- Darby, thomas. "what is biomass renewable energy". Real world energy. Archived from the original on 2014-06-08. Retrieved 12 june2014

- Disadvantages-of-biofuels: biofuel.org.uk, Retrieved 1 May, 2019

- Randor radakovits; robert e. Jinkerson; al darzins; matthew c. Posewitz1 (2010). "genetic engineering of algae for enhanced biofuel production". Eukaryotic cell. 9 (4): 486–501. Doi:10.1128/ec.00364-09. Pmc 2863401. Pmid 20139239

Generations of Biofuels $\quad$ **2**

- **First-generation Biofuels**

- **Second-generation Biofuels**

- **Third-generation Biofuels**

- **Fourth-generation Biofuels**

Biofuels have been divided into four generations on the basis of their sources, namely, first, second, third and fourth. These biofuels are derived from food crops, biomass, algae and electrofuels respectively. The topics elaborated in this chapter will help in gaining a better perspective about these generations of biofuels.

First-generation Biofuels

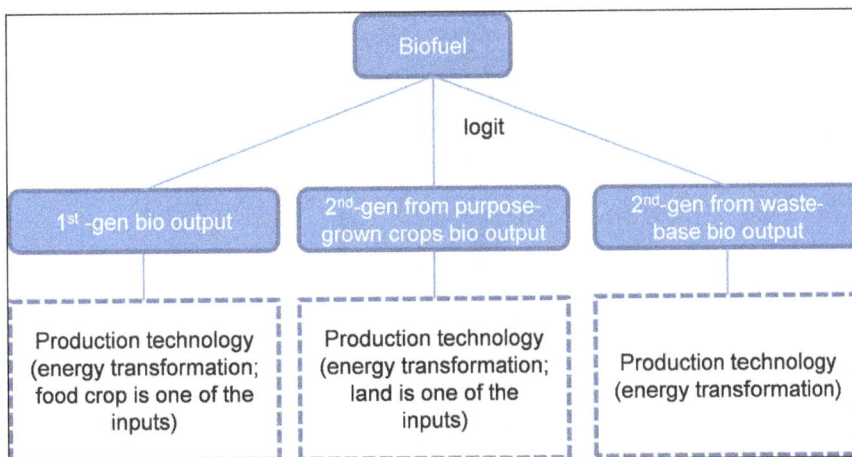

First Generation biofuels are produced directly from food crops by abstracting the oils for use in biodiesel or producing bioethanol through fermentation. Crops such as wheat and sugar are the most widely used feedstock for bioethanol while oil seed rape has proved a very effective crop for use in biodiesel. However, first generation biofuels have a number of associated problems. There is much debate over their actually benefit in reducing green house gas and co_2 emissions due to the fact that some biofuels can produce negative Net energy gains, releasing more carbon in their production than their feedstock's capture in their growth. However, the most contentious issue with first generation biofuels is 'fuel vs food'. As the majority of biofuels are produced directly from food crops the rise in demand for biofuels has lead to an increase in the volumes of crops being diverted away from the global food market. This has been blamed for the global increase in food prices over the last couple of years.

Second-generation Biofuels

Second G-eneration biofuels have been developed to overcome the limitations of first generation biofuels. They are produced from non-food crops such as wood, organic waste, food crop waste and specific biomass crops, therefore eliminating the main problem with first generation biofuels. Second Generation biofuels are also aimed at being more cost competitive in relation to existing fossil fuels. Life cycle assessments of second-generation biofuels have also indicated that they will increase 'net energy gains' over coming another of the main limitations of first generation biofuels.

Third-generation Biofuels

Solix Bioreactor.

The Third Generation of biofuels is based on improvements in the production of biomass. It takes advantage of specially engineered energy crops such as algae as its energy source. The algae are cultured to act as a low-cost, high-energy and entirely renewable feedstock. It is predicted that algae will have the potential to produce more energy per acre than conventional crops. Algae can also be grown using land and water unsuitable for food production, therefore reducing the strain on already depleted water sources. A further benefit of algae based biofuels is that the fuel can be manufactured into a wide range of fuels such as diesel, petrol and jet fuel.

Fourth-generation Biofuels

Four Generation Bio-fuels are aimed at not only producing sustainable energy but also a way of capturing and storing co2. Biomass materials, which have absorbed co2 while growing, are converted into fuel using the same processes as second generation biofuels. This process differs from second and third generation production as at all stages of production the carbon dioxide is captured using processes such as oxy-fuel combustion. The carbon dioxide can then be geosequestered by storing it in old oil and gas fields or saline aquifers. This carbon capture makes fourth generation biofuel production carbon negative rather then simply carbon neutral, as it is 'locks' away more carbon than it produces. This system not only captures and stores carbon dioxide from the atmosphere but it also reduces co2 emissions by replacing fossil fuels.

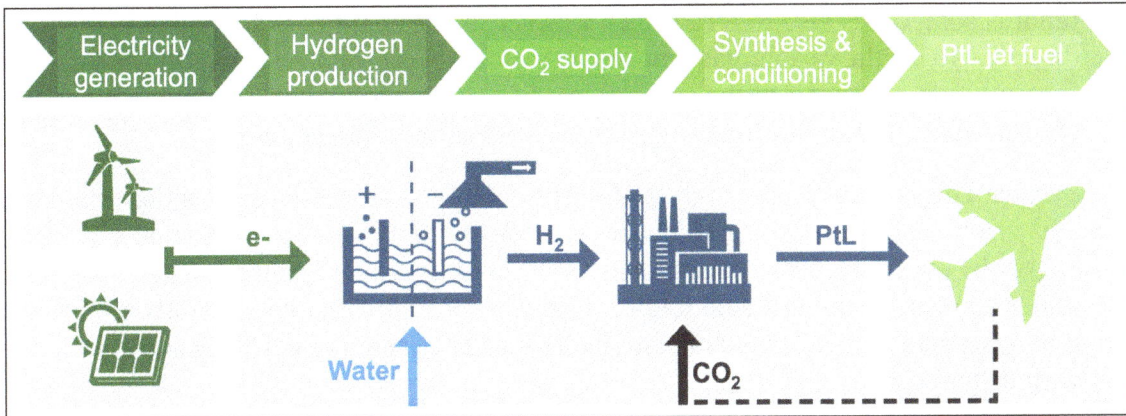

First-generation Biofuels

The first generation biofuels refer to the fuels that have been derived from sources like starch, sugar, animal fats and vegetable oil. The oil is obtained using the conventional techniques of production. Some of the most popular types of first generation biofuels are:

- Biodiesel: This is the most common type of biofuel commonly used in the European countries. This type of biofuel is mainly produced using a process called transesterification. This fuel if very similar to the mineral diesel and is chemically known as fatty acid methyl. This oil is produced after mixing the biomass with methanol and sodium hydroxide. The chemical reaction thereof produces biodiesel. Biodiesel is very commonly used for the various diesel engines after mixing up with mineral diesel. Now in many countries the manufacturers of the diesel engine ensure that the engine works well even with the biodiesel.

- Vegetable oil: These kinds of oil can be either used for cooking purpose or even as fuel. The main fact that determines the usage of this oil is the quality. The oil with good quality is generally used for cooking purpose. Vegetable oil can even be used in most of the old diesel engines, but only in warm atmosphere. In most of the countries, vegetable oil is mainly used for the production of biodiesel.

- Biogas: Biogas is mainly produced after the anaerobic digestion of the organic materials. Biogas can also be produced with the biodegradation of waste materials which are fed into anaerobic digesters which yields biogas. The residue or the by product can be easily used as manure or fertilizers for agricultural use. The biogas produced is very rich in methane which can be easily recovered through the use of mechanical biological treatment systems.

A less clean form of biogas is the landfill gas which is produced by the use of naturally occurring anaerobic digesters, but the main threat is that these gases can be a severe threat if escapes into the atmosphere.

- Bioalcohols: These are alcohols produced by the use if enzymes and micro organisms through the process of fermentation of starches and sugar. Ethanol is the most common type of bioalcohol whereas butanol and propanol are some of the lesser known ones. Biobutanol is sometimes also referred to as a direct replacement of gasoline because it can be directly used in the various gasoline engines. Butanol is produced using the process of ABE fermentation, and some of the experiments have also proved that butanol is a more energy efficient fuel and can be directly used in the various gasoline engines.

- Syngas: This is a gas that is produce after the combined process of gasification, combustion and pyrolysis. Biofuel used in this process is converted into carbon monoxide and then into energy by pyrolysis. During the process, very little oxygen is supplied to keep combustion under control. In the last step known as gasification the organic materials are converted into gases like carbon monoxide and hydrogen. The resulting gas Syngas can be used for various purposes.

First generation biofuels are produced directly from food crops. The biofuel is ultimately derived from the starch, sugar, animal fats, and vegetable oil that these crops provide. It is important to note that the structure of the biofuel itself does not change between generations, but rather the source from which the fuel is derived changes. Corn, wheat, and sugar cane are the most commonly used first generation biofuel feed stock.

Corn

Corn is the primary source of the world's fuel ethanol and most of that corn comes from the United States. As of 2012, more than 40 percent of the US corn crop was being used to produce corn ethanol, though not all of ethanol is used as biofuel. Current requirements by the United States government require that roughly 36 billion gallons of renewable biofuel be produced in 2013. Under the renewable fuel standard, up to 15 billion gallons of that will be grain based ethanol, including corn.

The Advantages of corn are:

- Infrastructure for planting, harvesting, and processing is already in place.

- Relatively simple conversion of corn starch to ethanol.

- Potential to use the rest of the plant (stalk, cob, etc.) to produce ethanol as well.

- Corn has the potential to supply about ¼ of U.S. gasoline consumption.

- There are no indirect land use costs with corn.

The Disadvantages of corn are:

- Relatively high requirement for pesticide and fertilizer. Not only is this expensive, but it leads to soil and water contamination.

- It is a food staple and use in biofuel has increased food prices worldwide, leading to hunger.

- The production rate is low at an average of just 350 gallons of fuel per acre.

- Energy yield is about 1.2, which is just barely positive at 20% net yield.

The general consensus seems to be the corn can never be anything more than a side show in the biofuel world. Its drawbacks, particularly it's important in the food chain, prevent corn from being a viable alternative fuel feedstock.

Sugar Cane

Not far behind corn in terms of overall ethanol production is sugar cane. The majority of the world's sugar cane is grown in Brazil, which was the world's largest producer of alcohol fuel until very recently went it was eclipsed by the United States. Brazil produces roughly 5 billion gallons or 18 billion litres of fuel ethanol annually. The country adopted a very favourable stance on ethanol derived from sugar cane as a result of the oil embargo of the 1970s. Brazil has a policy of at least 22% ethanol in its gasoline, though 100% ethanol is available for purchase.

Unlike corn, sugar cane provides sugar rather than starch, which is more easily converted to alcohol. Where as corn requires heating and then fermentation, sugar cane requires only fermentation.

The Advantages of sugar cane include:

- Infrastructure for planting, harvesting, and processing that is already in place.

- No land use changes provide plantations sizes remain stable.

- The yield is higher than that of corn at an average of 650 gallons per acre.

- Carbon dioxide emissions can be 90% lower than for conventional gasoline when land use changes do not occur.

The Disadvantages of sugar cane include:

- Despite having a higher yield than corn, it is still relatively low.

- Few regions are suitable to cultivation.

- Sugar cane is a food staple in countries of South and Central America.

Like corn, sugar cane is not considered a viable solution to the world's energy needs. It suits Brazil and a few other countries well, but cannot be scaled for a number of reasons.

Soybeans

Unlike corn and sugar cane, soybeans are grown throughout much of North America, South America, and Asia. In other words, soybeans are a global food crop. The United States produce roughly 32 percent of all soybeans in the world, followed by Brazil at 28 percent. Despite its relatively high price as a food crop, soybean is still a major feedstock for the production of biofuel. In this case, rather than ethanol, soybean is used to produce biodiesel. Soybean is probably the worst feedstock for biofuel production.

The Advantages of soybeans include:

- Grows in many regions.

- Relatively easy to maintain.

The Disadvantages of soybeans include:

- A yield of only about 70 gallons of biodiesel per acre, which is the worst yield of any crop. Palm oil produces almost 10 times as much biodiesel per acre at 600 gallons (note palm oil is considered a second generation feedstock).

- Soybean is a common food source and thus its use as a biofuel directly threatens the food chain.

- It faces a number of disease and pest burdens.

- It is generally not a profitable biofuel feedstock.

- More energy is usually required to cultivate soybeans than can be derived from the fuel produced from them.

Vegetable Oil

Vegetable oil, which can be derived from any number of vegetables, can fall into the category of both a first and a second generation biofuel. If used directly as "virgin" vegetable oil, it is a first generation biofuel. If used after it is no longer fit for cooking, then vegetable oil becomes a second generation biofuel. Here we consider only its benefits and drawbacks as a first generation feedstock used in the production of biodiesel.

The Advantages of vegetable oil:

- It is easy to convert to biodiesel.

- It is widely available.

- It can often be used directly in diesel engines with little modification.

The Disadvantages of vegetable oil:

- It is an important feedstock.

- When unrefined, it can cause engine damage through carbon deposition due to incomplete combustion.

- The replacement of old growth forest with oil palms increases carbon emissions and damages biodiversity.

Other Candidate Crops

Wheat, sugar beets, rapeseed, peanuts, and a number of other food crops have all, at one point or another, served as feedstock for biofuel. However, they all suffer from the same problems including threatening the food chain, increasing carbon emissions when planted outside traditional agricultural settings, and intense growth requirements. Ultimately, first generation biofuels have given way to second and third generation fuels for the reasons mentioned above. Though first generation feedstock will provide biofuel for the foreseeable future, their importance is waning and new, better alternatives are being developed.

Second-generation Biofuels

Second-generation biofuels, also known as advanced biofuels, are fuels that can be manufactured from various types of non-food biomass. Biomass in this context means plant materials and animal waste used especially as a source of fuel.

First-generation biofuels are made from the sugars and vegetable oils found in food crops using standard processing technologies. Second-generation biofuels are made from different feedstocks and therefore may require different technology to extract useful energy from them. Second generation feedstocks include lignocellulosic biomass or woody crops, agricultural residues or waste, as well as dedicated non-food energy crops grown on marginal land unsuitable for food production.

The term second-generation biofuels is used loosely to describe both the 'advanced' technology used to process feedstocks into biofuel, but also the use of non-food crops, biomass and wastes as feedstocks in 'standard' biofuels processing technologies if suitable. This causes some considerable confusion. Therefore it is important to distinguish between second-generation feedstocks and second-generation biofuel processing technologies.

The development of second-generation biofuels has seen a stimulus since the food vs. fuel dilemma regarding the risk of diverting farmland or crops for biofuels production to the detriment of food supply. The biofuel and food price debate involves wide-ranging views, and is a long-standing, controversial one in the literature.

Second-generation biofuel technologies have been developed to enable the use of non-food biofuel feedstocks because of concerns to food security caused by the use of food crops for the production of first-generation biofuels. The diversion of edible food biomass to the production of biofuels could theoretically result in competition with food and land uses for food crops.

First-generation bioethanol is produced by fermenting plant-derived sugars to ethanol, using a similar process to that used in beer and wine-making. This requires the use of food and fodder crops, such as sugar cane, corn, wheat, and sugar beet. The concern is that if these food crops are used for biofuel production that food prices could rise and shortages might be experienced in some countries. Corn, wheat, and sugar beet can also require high agricultural inputs in the form of fertilizers, which limit the greenhouse gas reductions that can be achieved. Biodiesel produced by transesterification from rapeseed oil, palm oil, or other plant oils is also considered a first-generation biofuel.

The goal of second-generation biofuel processes is to extend the amount of biofuel that can be produced sustainably by using biomass consisting of the residual non-food parts of current crops, such as stems, leaves and husks that are left behind once the food crop has been extracted, as well as other crops that are not used for food purposes (non-food crops), such as switchgrass, grass, jatropha, whole crop maize, miscanthus and cereals that bear little grain, and also industry waste such as woodchips, skins and pulp from fruit pressing, etc.

The problem that second-generation biofuel processes are addressing is to extract useful feed-stocks from this woody or fibrous biomass, where the useful sugars are locked in by lignin, hemi-cellulose and cellulose. All plants contain lignin, hemicellulose and cellulose. These are complex carbohydrates (molecules based on sugar). Lignocellulosic ethanol is made by freeing the sugar molecules from cellulose using enzymes, steam heating, or other pre-treatments. These sugars can then be fermented to produce ethanol in the same way as first-generation bioethanol pro-duction. The by-product of this process is lignin. Lignin can be burned as a carbon neutral fuel to produce heat and power for the processing plant and possibly for surrounding homes and businesses. Thermochemical processes (liquefaction) in hydrothermal media can produce liquid oily products from a wide range of feedstock that has a potential to replace or augment fuels. However, these liquid products fall short of diesel or biodiesel standards. Upgrading liquefac-tion products through one or many physical or chemical processes may improve properties for use as fuel.

Second-generation Technology

The following subsections describe the main second-generation routes currently under develop-ment.

Thermochemical Routes

Carbon-based materials can be heated at high temperatures in the absence (pyrolysis) or presence of oxygen, air and/or steam (gasification).

These thermochemical processes yield a mixture of gases including hydrogen, carbon monoxide, carbon dioxide, methane and other hydrocarbons, and water. Pyrolysis also produces a solid char. The gas can be fermented or chemically synthesised into a range of fuels, including ethanol, syn-thetic diesel, synthetic gasoline or jet fuel.

There are also lower temperature processes in the region of 150–374 °C, that produce sugars by decomposing the biomass in water with or without additives.

Gasificatio

Gasification technologies are well established for conventional feedstocks such as coal and crude oil. Second-generation gasification technologies include gasification of forest and agricultural residues, waste wood, energy crops and black liquor. Output is normally syngas for further synthesis to e.g. Fischer-Tropsch products including diesel fuel, biomethanol, BioDME (dimethyl ether), gasoline via catalytic conversion of dimethyl ether, or biomethane (synthetic natural gas). Syngas can also be used in heat production and for generation of mechanical and electrical power via gas motors or gas turbines.

Pyrolysis

Pyrolysis is a well established technique for decomposition of organic material at elevated temperatures in the absence of oxygen. In second-generation biofuels applications forest and agricultural residues, wood waste and energy crops can be used as feedstock to produce e.g. bio-oil for fuel oil applications. Bio-oil typically requires significant additional treatment to render it suitable as a refinery feedstock to replace crude oil.

Torrefaction

Torrefaction is a form of pyrolysis at temperatures typically ranging between 200–320 °C. Feedstocks and output are the same as for pyrolysis.

Hydrothermal Liquefaction

Hydrothermal liquefaction is a process similar to pyrolysis that can process wet materials. The process is typically at moderate temperatures up to 400 °C and higher than atmospheric pressures. The capability to handle a wide range of materials make hydrothermal liquefaction viable for producing fuel and chemical production feedstock.

Biochemical Routes

Chemical and biological processes that are currently used in other applications are being adapted for second-generation biofuels. Biochemical processes typically employ pre-treatment to accelerate the hydrolysis process, which separates out the lignin, hemicellulose and cellulose. Once these ingredients are separated, the cellulose fractions can be fermented into alcohols.

Feedstocks are energy crops, agricultural and forest residues, food industry and municipal biowaste and other biomass containing sugars. Products include alcohols (such as ethanol and butanol) and other hydrocarbons for transportation use.

Types of Second-generation Biofuels

The following second-generation biofuels are under development, although most or all of these biofuels are synthesized from intermediary products such as syngas using methods that are identical in processes involving conventional feedstocks, first-generation and second-generation biofuels. The distinguishing feature is the technology involved in producing the intermediary product, rather than the ultimate off-take.

A process producing liquid fuels from gas (normally syngas) is called a gas-to-liquid (GtL) process. When biomass is the source of the gas production the process is also referred to as biomass-to-liquids (BTL).

From Syngas using Catalysis

- Biomethanol can be used in methanol motors or blended with petrol up to 10–20% without any infrastructure changes.

- BioDME can be produced from Biomethanol using catalytic dehydration or it can be produced directly from syngas using direct DME synthesis. DME can be used in the compression ignition engine.

- Bio-derived gasoline can be produced from DME via high-pressure catalytic condensation reaction. Bio-derived gasoline is chemically indistinguishable from petroleum-derived gasoline and thus can be blended into the gasoline pool.

- Biohydrogen can be used in fuel cells to produce electricity.

- Mixed Alcohols (i.e., mixture of mostly ethanol, propanol, and butanol, with some pentanol, hexanol, heptanol, and octanol). Mixed alcohols are produced from syngas with several classes of catalysts. Some have employed catalysts similar to those used for methanol. Molybdenum sulfide catalysts were discovered at Dow Chemical and have received considerable attention. Addition of cobalt sulfide to the catalyst formulation was shown to enhance performance. Molybdenum sulfide catalysts have been well studied but have yet to find widespread use. These catalysts have been a focus of efforts at the U.S. Department of Energy's Biomass Program in the Thermochemical Platform. Noble metal catalysts have also been shown to produce mixed alcohols. Most R&D in this area is concentrated in producing mostly ethanol. However, some fuels are marketed as mixed alcohols Mixed alcohols are superior to pure methanol or ethanol, in that the higher alcohols have higher energy content. Also, when blending, the higher alcohols increase compatibility of gasoline and ethanol, which increases water tolerance and decreases evaporative emissions. In addition, higher alcohols have also lower heat of vaporization than ethanol, which is important for cold starts. Biomethane (or Bio-SNG) via the Sabatier reaction.

From Syngas using Fischer–Tropsch

The Fischer–Tropsch (FT) process is a gas-to-liquid (GtL) process. When biomass is the source of the gas production the process is also referred to as biomass-to-liquids (BTL). A disadvantage of this process is the high energy investment for the FT synthesis and consequently, the process is not yet economic.

FT diesel can be mixed with fossil diesel at any percentage without need for infrastructure change and moreover, synthetic kerosene can be produced.

Biocatalysis

- Biohydrogen might be accomplished with some organisms that produce hydrogen directly under certain conditions. Biohydrogen can be used in fuel cells to produce electricity.

- Butanol and Isobutanol via recombinant pathways expressed in hosts such as E. coli and yeast, butanol and isobutanol may be significant products of fermentation using glucose as a carbon and energy source.

- DMF (2,5-Dimethylfuran). Recent advances in producing DMF from fructose and glucose using catalytic biomass-to-liquid process have increased its attractiveness.

Other Processes

- HTU (Hydro Thermal Upgrading) diesel is produced from wet biomass. It can be mixed with fossil diesel in any percentage without need for infrastructure.

- Wood diesel. A new biofuel was developed by the University of Georgia from woodchips. The oil is extracted and then added to unmodified diesel engines. Either new plants are used or planted to replace the old plants. The charcoal byproduct is put back into the soil as a fertilizer. According to the director Tom Adams since carbon is put back into the soil, this biofuel can actually be carbon negative not just carbon neutral. Carbon negative decreases carbon dioxide in the air reversing the greenhouse effect not just reducing it.

Second-generation Feedstocks

To qualify as a second generation feedstock, a source must not be suitable for human consumption. Second-generation biofuel feedstocks include specifically grown inedible energy crops, cultivated inedible oils, agricultural and municipal wastes, waste oils, and algae. Nevertheless, cereal and sugar crops are also used as feedstocks to second-generation processing technologies. Land use, existing biomass industries and relevant conversion technologies must be considered when evaluating suitability of developing biomass as feedstock for energy.

Energy Crops

Plants are made from lignin, hemicellulose and cellulose; second-generation technology uses one, two or all of these components. Common lignocellulosic energy crops include wheat straw, Arundo donax, Miscanthus spp., short rotation coppice poplar and willow. However, each offers different opportunities and no one crop can be considered 'best' or 'worst'.

Municipal Solid Waste

Municipal Solid Waste comprises a very large range of materials, and total waste arisings are increasing. In the UK, recycling initiatives decrease the proportion of waste going straight for disposal, and the level of recycling is increasing each year. However, there remains significant opportunities to convert this waste to fuel via gasification or pyrolysis.

Green Waste

Green waste such as forest residues or garden or park waste may be used to produce biofuel via different routes. Examples include Biogas captured from biodegradable green waste, and gasification or hydrolysis to syngas for further processing to biofuels via catalytic processes.

Black Liquor

Black liquor, the spent cooking liquor from the kraft process that contains concentrated lignin and hemicellulose, may be gasified with very high conversion efficiency and greenhouse gas reduction potential to produce syngas for further synthesis to e.g. biomethanol or BioDME.

The yield of crude tall oil from process is in the range of 30 – 50 kg / ton pulp.

Greenhouse Gas Emissions

Lignocellulosic biofuels reduces greenhouse gas emissions by 60–90% when compared with fossil petroleum which is on par with the better of current biofuels of the first-generation, where typical best values currently is 60–80%. In 2010, average savings of biofuels used within EU was 60%. In 2013, 70% of the biofuels used in Sweden reduced emissions with 66% or higher.

Commercial Development

An operating lignocellulosic ethanol production plant is located in Canada, run by Iogen Corporation. The demonstration-scale plant produces around 700,000 litres of bioethanol each year. A commercial plant is under construction. Many further lignocellulosic ethanol plants have been proposed in North America and around the world.

The Swedish specialty cellulose mill Domsjö Fabriker in Örnsköldsvik, Sweden develops a biorefinery using Chemrec's black liquor gasification technology. When commissioned in 2015 the biorefinery will produce 140,000 tons of biomethanol or 100,000 tons of BioDME per year, replacing 2% of Sweden's imports of diesel fuel for transportation purposes. In May 2012 it was revealed that Domsjö pulled out of the project, effectively killing the effort.

In the UK, companies like INEOS Bio and British Airways are developing advanced biofuel refineries, which are due to be built by 2013 and 2014 respectively. Under favourable economic conditions and strong improvements in policy support, NNFCC projections suggest advanced biofuels could meet up to 4.3 per cent of the UK's transport fuel by 2020 and save 3.2 million tonnes of CO_2 each year, equivalent to taking nearly a million cars off the road.

Helsinki, Finland, 1 February 2012 – UPM is to invest in a biorefinery producing biofuels from crude tall oil in Lappeenranta, Finland. The industrial scale investment is the first of its kind globally. The biorefinery will produce annually approximately 100,000 tonnes of advanced second-generation biodiesel for transport. Construction of the biorefinery will begin in the summer of 2012 at UPM's Kaukas mill site and be completed in 2014. UPM's total investment will amount to approximately EUR 150 million.

Calgary, Alberta, 30 April 2012 – Iogen Energy Corporation has agreed to a new plan with its joint owners Royal Dutch Shell and Iogen Corporation to refocus its strategy and activities. Shell continues to explore multiple pathways to find a commercial solution for the production of advanced biofuels on an industrial scale, but the company will NOT pursue the project it has had under development to build a larger scale cellulosic ethanol facility in southern Manitoba.

Third-generation Biofuels

The term third generation biofuel has only recently enter the mainstream it refers to biofuel derived from algae. Previously, algae were lumped in with second generation biofuels. However, when it became apparent that algae are capable of much higher yields with lower resource inputs than other feedstock, many suggested that they be moved to their own category. Algae provide a number of advantages, but at least one major shortcoming that has prevented them from becoming a runaway success.

Fuel Potential of Third-generation Biofuels

When it comes to the potential to produce fuel, no feedstock can match algae In terms of quantity or diversity. The diversity of fuel that algae can produce results from two characteristics of the microorganism. First, algae produce an oil that can easily be refined into diesel or even certain components of gasoline. More importantly, however, is a second property in it can be genetically manipulated to produce everything from ethanol and butanol to even gasoline and diesel fuel directly.

Butanol is of great interest because the alcohol is exceptionally similar to gasoline. In fact, it has a nearly identical energy density to gasoline and an improved emissions profile. Until the advent of genetically modified algae, scientists had a great deal of difficulty producing butanol. Now, several commercial-scale facilities have been developed and are on the brink of making butanol and more popular biofuel than ethanol because it is not only similar in many ways to gasoline, but also does not cause engine damage or even require engine modification the way ethanol does.

The list of fuels that can be derived from algae includes:

- Biodiesel
- Butanol
- Gasoline
- Methane
- Ethanol
- Vegetable Oil
- Jet Fuel

Diversity is not the only thing that algae has going for it in terms of fuel potential. It is also capable of producing outstanding yields. In fact, algae have been used to produce up to 9000 gallons of biofuel per acre, which is 10-fold what the best traditional feedstock have been able to generate. People who work closely with algae have suggested that yields as high as 20,000 gallons per acre are attainable. According to the US Department of Energy, yields that are 10 times higher than second generation biofuels mean that only 0.42% of the U.S. land area would be needed to generate enough biofuel to meet all of the U.S. needs. Given that the U.S. is the largest consumer of fuel in the world, that is saying something about the efficiency of algal-based biofuels.

Cultivation of Third-generation Biofuels

Another favorable property of algae is the diversity of ways in which it can be cultivated. Algae can be grown in any of the following ways:

- Open ponds – These are the simplest systems in which algae is grown in a pond in the open air. They are simple and have low capital costs, but are less efficient than other systems. They are also of concern because other organisms can contaminate the pond and potentially damage or kill the algae.
- Closed-loop systems – These are similar to open ponds, but they are not exposed to the

atmosphere and use a sterile source of carbon dioxide. Such systems have potential because they may be able to be directly connected to carbon dioxide sources (such as smokestacks) and thus use the gas before it is every released into the atmosphere.

- Photobioreactors – These are the most advanced and thus most difficult systems to implement, resulting in high capital costs. Their advantages in terms of yield and control, however, are unparalleled. They are closed systems.

Note that all three systems mean that algae are able to be grown almost anywhere that temperatures are warm enough. This means that no farm land need be threatened by algae. Closed-loop and photobioreactor systems have even been used in desert settings.

What is more, algae can be grown in waste water, which means they can offer secondary benefits by helping to digest municipal waste while avoiding taking up any additional land. All of the factors above combine to make algae easier to cultivate than traditional biofuels.

Third-generation Biofuel Feedstock

One of the major benefits of algae is that they can use a diverse array of carbon sources. Most notably, it has been suggested that algae might be tied directly to carbon emitting sources (power plants, industry, etc.) where they could directly convert emissions into usable fuel. This means that no carbon dioxide would be released from these settings and thus total emissions would be reduced substantially.

As with everything, algae have a down side. In this case, the downside is large and if it cannot be solved, is a deal breaker. Algae, even when grown in waste water, require large amounts of water, nitrogen and phosphorus to grow. So much in fact that the production of fertilizer to meet the needs of algae used to produce biofuel would produce more greenhouse gas emissions than were saved by using algae based biofuel to begin with. It also means the cost of algae-base biofuel is much higher than fuel from other sources.

This single disadvantage means that the large-scale implementation of algae to produce biofuel will not occur for a long time, if at all. In fact, after investing more than $600 million USD into research and development of algae, Exxon Mobil came to the conclusion in 2013 that algae-based biofuels will not be viable for at least 25 years. What is more, that calculation is strictly economical and does not consider the environmental impacts that have yet to be solved.

A minor drawback regarding algae is that biofuel produced from them tends to be less stable than biodiesel produced from other sources. This is because the oil found in algae tends to be highly unsaturated. Unsaturated oils are more volatile, particularly at high temperatures, and thus more prone to degradation. Unlike the fertilizer requirements above, this is a problem that has a potential solution.

Fourth-generation Biofuels

Fourth generation biofuels are derived from specially engineered plants or biomass that may have higher energy yields or lower barriers to cellulosic breakdown or are able to be grown on non-agricultural land or bodies of water.

In fourth generation production systems, biomass crops are seen as efficient 'carbon capturing' machines that take CO2 out of the atmosphere and 'lock' it in their branches, trunks and leaves. Then, the carbon-rich biomass is converted into fuel and gases by means of second generation techniques. Crucially, before, during or after the bioconversion process, the carbon dioxide is captured by utilizing so-called pre-combustion, oxyfuel or post-combustion processes. The greenhouse gas is then geosequestered - stored in depleted oil and gas fields, in unmineable coal seams or in saline aquifers, where it stays locked up for hundreds, possibly thousands of years.

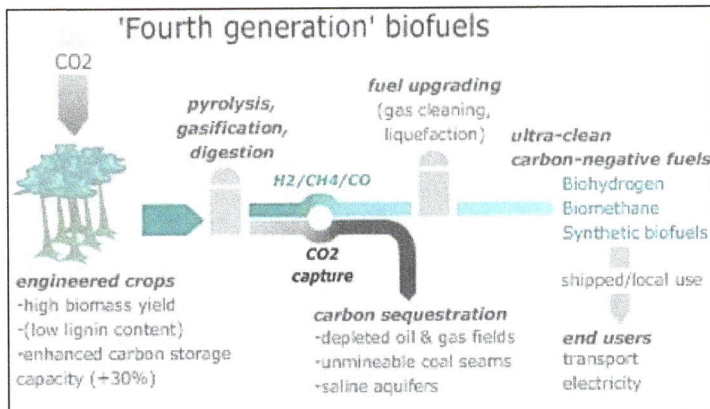

The resulting fuels and gases are not only renewable, they are also effectively carbon-negative. Only the utilization of biomass allows for the conception of carbon-negative energy; all other renewables (wind, solar, etc) are all carbon-neutral at best, carbon-positive in practice. Fourth generation biofuels instead take historic CO_2 emissions out of the atmosphere. The system not only captures and stores carbon dioxide from the atmosphere, it also reduces carbon dioxide emission by replacing fossil fuels.

Electrofuel

Electrofuels (synthetic fuel) are an emerging class of carbon-neutral drop-in replacement fuels that are made by storing electrical energy from renewable sources in the chemical bonds of liquid or gas fuels. The primary targets are butanol, biodiesel, and hydrogen, but include other alcohols and carbon-containing gasses such as methane and butane.

A primary source of funding for research on liquid electrofuels for transportation is the Electrofuels Program of the Advanced Research Projects Agency-Energy (ARPA-E), headed by Eric Toone. ARPA-E, created in 2009 under President Obama's Secretary of Energy Steven Chu, is the Department of Energy's (DOE) attempt to duplicate the effectiveness of the Defense Advanced Research Projects Agency, DARPA. Examples of projects funded under this program include OPX Biotechnologies' biodiesel effort led by Michael Lynch and Derek Lovley's work on microbial electrosynthesis at the University of Massachusetts Amherst, which reportedly produced the first liquid electrofuel using CO_2 as the feedstock.

The first Electrofuels Conference, sponsored by the American Institute of Chemical Engineers was held in Providence, RI in November 2011. At that conference, Director Eric Toone stated that "Eighteen months into the program, we know it works. We need to know if we can make it matter." Several groups are beyond proof-of-principle, and are working to scale up cost-effectively.

Electrofuels have the potential to be disruptive if carbon-neutral electrofuels can be cheaper than petroleum fuels, and chemical feedstocks produced by electrosynthesis cheaper than those refined from crude oil. Electrofuels also have a great potential to alter the renewable energy landscape, as electrofuels allow renewables from all sources to be stored conveniently as a liquid fuel.

As of 2014, prompted by the fracking boom, ARPA-E's focus has moved from electrical feedstocks to natural-gas based feedstocks, and thus away from electrofuels.

Examples: Audi is working on E-diesel and E-gasoline.

Solar Fuel

A solar fuel is a synthetic chemical fuel produced directly/indirectly from solar energy sunlight/ solar heat through photochemical/photobiological, thermochemical (i.e., through the use of solar heat supplied by concentrated solar thermal energy to drive a chemical reaction), and electrochemical reaction. Light is used as an energy source, with solar energy being transduced to chemical energy, typically by reducing protons to hydrogen, or carbon dioxide to organic compounds. A solar fuel can be produced and stored for later usage, when sunlight is not available, making it an alternative to fossil fuels. Diverse photocatalysts are being developed to carry these reactions in a sustainable, environmentally friendly way.

The world's dependence on the declining reserves of fossil fuels poses not only environmental problems but also geopolitical ones. Solar fuels, in particular hydrogen, are viewed as an alternative source of energy for replacing fossil fuels especially where storage is essential. Electricity can be produced directly from sunlight through photovoltaics, but this form of energy is rather inefficient to store compared to hydrogen. A solar fuel can be produced when and where sunlight is available, and stored and transported for later usage.

The most widely researched solar fuels are hydrogen and products of carbon dioxide reduction.

Solar fuels can be produced via direct or indirect processes. Direct processes harness the energy in sunlight to produce a fuel without intermediary energy conversions. In contrast, indirect processes have solar energy converted to another form of energy first (such as biomass or electricity) that can then be used to produce a fuel. Indirect processes have been easier to implement but have the disadvantage of being less efficient than, e.g., water splitting for the production of hydrogen, since energy is wasted in the intermediary conversion.

Photochemical Hydrogen Production

Hydrogen can be produced by electrolysis. To use sunlight in this process, a photoelectrochemical cell can be used, where one photosensitized electrode converts light into an electric current that is then used for water splitting. One such type of cell is the dye-sensitized solar cell. This is an indirect process, since it produces electricity that then is used to form hydrogen. The other major indirect process using sunlight is conversion of biomass to biofuel using photosynthetic organisms; however, most of the energy harvested by photosynthesis is used in life-sustaining processes and therefore lost for energy use.

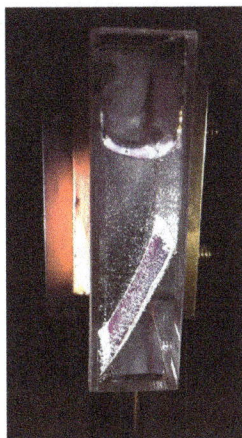

A sample of a photoelectric cell in a lab environment. Catalysts are added to the cell, which is submerged in water and illuminated by simulated sunlight. The bubbles seen are oxygen (forming on the front of the cell) and hydrogen (forming on the back of the cell).

A direct process can use a catalyst that reduces protons to molecular hydrogen upon electrons from an excited photosensitizer. Several such catalysts have been developed as proof of concept, but not yet scaled up for commercial use; nevertheless, their relative simplicity gives the advantage of potential lower cost and increased energy conversion efficiency. One such proof of concept is the "artificial leaf" developed by Nocera and coworkers: a combination of metal oxide-based catalysts and a semiconductor solar cell produces hydrogen upon illumination, with oxygen as the only by-product.

Hydrogen can also be produced from some photosynthetic microorganisms (microalgae and cyanobacteria) using photobioreactors. Some of these organisms produce hydrogen upon switching culture conditions; for example, *Chlamydomonas reinhardtii* produces hydrogen anaerobically under sulfur deprivation, that is, when cells are moved from one growth medium to another that does not contain sulfur, and are grown without access to atmospheric oxygen. Another approach was to abolish activity of the hydrogen-oxidizing (uptake) hydrogenase enzyme in the diazotrophic cyanobacterium *Nostoc punctiforme*, so that it would not consume hydrogen that is naturally produced by the nitrogenase enzyme in nitrogen-fixing conditions. This *N. punctiforme* mutant could then produce hydrogen when illuminated with visible light.

Photochemical Carbon Dioxide Reduction

Carbon dioxide (CO_2) can be reduced to carbon monoxide (CO) and other more reduced compounds, such as methane, using the appropriate photocatalysts. One early example was the use of Tris(bipyridine)ruthenium(II) chloride ($Ru(bipy)_3Cl_2$) and cobalt chloride ($CoCl_2$) for CO_2 reduction to CO. Many compounds that do similar reactions have since been developed, but they generally perform poorly with atmospheric concentrations of CO_2, requiring further concentration. The simplest product from CO_2 reduction is carbon monoxide (CO), but for fuel development, further reduction is needed, and a key step also needing further development is the transfer of hydride anions to CO.

Also in this case, the use of microorganisms has been explored. Using genetic engineering and synthetic biology techniques, parts of or whole biofuel-producing metabolic pathways can be introduced in photosynthetic organisms. One example is the production of 1-butanol in *Synechococcus elongatus* using enzymes from *Clostridium acetobutylicum*, *Escherichia coli* and *Treponema denticola*. One example of a large-scale research facility exploring this type of biofuel production is the AlgaePARC in the Wageningen University and Research Centre, Netherlands.

Other Applications

- Electrolysis of water for hydrogen production combined with Solar PV using alkaline, PEM, and SOEC electrolyzers.

- Electro-catalytic CO_2 conversion using electrochemical reduction of CO_2, UV light photolysis, metal oxide based photocatalytic reduction of CO_2, and thermochemical reduction at high temperature.

References

- Generations-of-biofuels: energyfromwasteandwood.weebly.com, Retrieved 18 June, 2019

- Evans, G. "International Biofuels Strategy Project. Liquid Transport Biofuels - Technology Status Report, NNFCC 08-017", National Non-Food Crops Centre, 2008-04-14. Retrieved on 2011-02-16

- Oliver R. Inderwildi, David A. King (2009). "Quo Vadis Biofuels". Energy & Environmental Science. 2 (4): 343. Doi:10.1039/b822951c

- Third-generation-biofuels: biofuel.org.uk, Retrieved 10 August, 2019

- National Non-Food Crops Centre. "NNFCC Newsletter – Issue 19. Advanced Biofuels", Retrieved on 2011-06-27

- Fourth-generation, biofuels: wixsite.com, Retrieved 5 May, 2019

- "ELECTROFUELS: Microorganisms for Liquid Transportation Fuel". ARPA-E. Retrieved July 23, 2013

- First-generation-biofuels: biofuel.org.uk, Retrieved 13 July, 2019

- "Novel Biological Conversion of Hydrogen and Carbon Dioxide Directly into Free Fatty Acids". ARPA-E. Archived from the original on October 10, 2013. Retrieved July 23, 2013

Types of Biofuels 3

- **Biogas**

- **Syngas**

- **Alcohol Fuel**

- **Algae Fuel**

- **Biodiesel**

- **Vegetable Oil Fuel**

- **Aviation Biofuel**

- **Biogasoline**

- **Biohydrogen**

- **Bioethers**

- **Sustainable Biofuel**

Some of the common types of biofuels are biogas, syngas, algae fuel , biodiesel, vegetable oil fuel, biohydrogen, bioethers, alcohol fuel, aviation biofuel and biogasoline. This chapter has been carefully written to provide an easy understanding of these types of biofuels.

Biogas

Biogas is a type of biofuel that is naturally produced from the decomposition of organic waste. When organic matter, such as food scraps and animal waste, break down in an anaerobic environment (an environment absent of oxygen) they release a blend of gases, primarily methane and carbon dioxide. Because this decomposition happens in an anaerobic environment, the process of producing biogas is also known as anaerobic digestion.

Anaerobic digestion is a natural form of waste-to-energy that uses the process of fermentation to breakdown organic matter. Animal manure, food scraps, wastewater, and sewage are all examples of organic matter that can produce biogas by anaerobic digestion. Due to the high content of methane in biogas (typically 50-75%) biogas is flammable, and therefore produces a deep blue flame, and can be used as an energy source.

The Ecology of Biogas

Biogas is known as an environmentally-friendly energy source because it alleviates two major environmental problems simultaneously:

1. The global waste epidemic that releases dangerous levels of methane gas every day.

2. The reliance on fossil fuel energy to meet global energy demand.

By converting organic waste into energy, biogas is utilizing nature's elegant tendency to recycle substances into productive resources. Biogas generation recovers waste materials that would otherwise pollute landfills; prevents the use of toxic chemicals in sewage treatment plants, and saves money, energy, and material by treating waste on-site. Moreover, biogas usage does not require fossil fuel extraction to produce energy.

Instead, biogas takes a problematic gas, and converts it into a much safer form. More specifically, the methane content present in decomposing waste is converted into carbon dioxide. Methane gas has approximately 20 to 30 times the heat-trapping capabilities of carbon dioxide. This means that when a rotting loaf of bread converts into biogas, the loaf's environmental impact will be about 10 times less potent than if it was left to rot in a landfill.

Biogas Digesters

As opposed to letting methane gas release to the atmosphere, biogas digesters are the systems that process waste into biogas, and then channel that biogas so that the energy can be productively used. There are several types of biogas systems and plants that have been designed to make efficient use of biogas. While each model differs depending on input, output, size, and type, the biological process that converts organic waste into biogas is uniform. Biogas digesters receive organic matter, which decompose in a digestion chamber. The digestion chamber is fully submerged in water, making it an anaerobic (oxygen-free) environment. The anaerobic environment allows for microorganisms to break down the organic material, and convert it into biogas.

All-Natural Fertilizer

Because the organic material decomposes in a liquid environment, nutrients present in the waste dissolve into the water, and create a nutrient-rich sludge, typically used as fertilizer for plants. This fertilizer output is generated on a daily basis, and therefore is a highly productive by-product of anaerobic digestion.

Biological Breakdown

To produce biogas, organic matter ferments with the help of bacterial communities. Four stages of fermentation move the organic material from their initial composition into their biogas state.

1. The first stage of the digestion process is the hydrolysis stage. In the hydrolysis stage insoluble organic polymers (such as carbohydrates) are broken down, making it accessible to the next stage of bacteria called acidogenic bacteria.

2. The acideogenic bacteria convert sugars and amino acids into carbon dioxide, hydrogen, ammonia, and organic acids.

3. At the third stage the acetogenic bacteria convert the organic acids into acetic acid, hydrogen, ammonia, and carbon dioxide, allowing for the final stage- the methanogens.

4. The methanogens convert these final components into methane and carbon dioxide- which can then be used as a flammable, green energy.

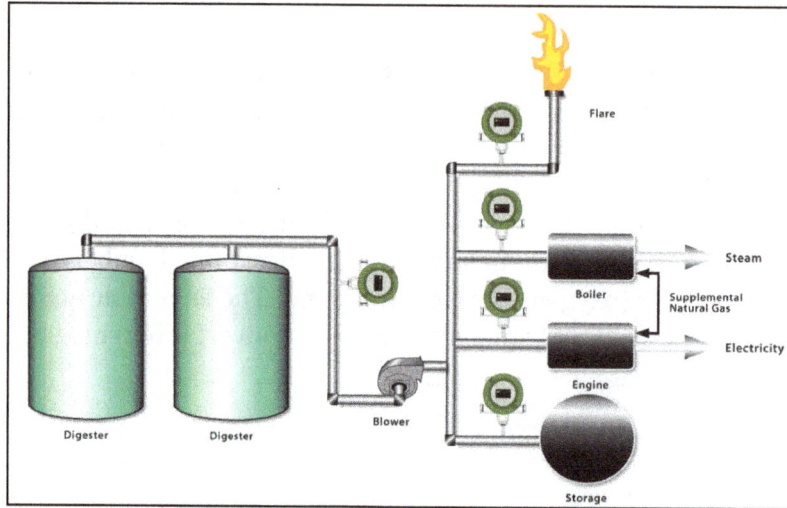

Many uses of Biogas

Biogas can be produced with various types of organic matter, and therefore there are several types of models for biogas digesters. Some industrial systems are designed to treat: municipal wastewater, industrial wastewater, municipal solid waste, and agricultural waste.

Small-scale systems are typically used for digesting animal waste. And newer family-size systems are designed to digest food waste. The resulting biogas can be used in several ways including: gas, electricity, heat, and transportation fuels.

For example, in Sweden hundreds of cars and buses run on refined biogas. The biogas in Sweden is produced primarily from sewage treatment plants and landfills.

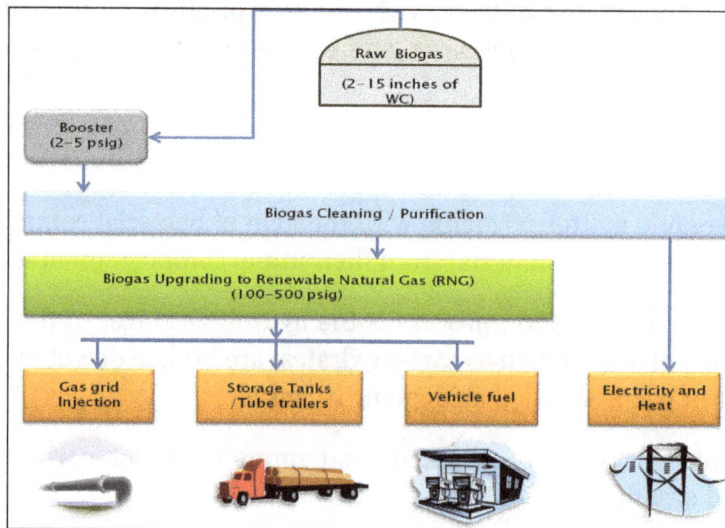

Another example of the diversified uses of biogas is the First Milk plant. One of the UK's biggest cheese makers is building an anaerobic digestion plant that will process dairy residues and convert into bio-methane for the gas grid. New anaerobic digestion plants like these with fascinating stories keep popping up every day.

Small-scale Biogas Systems

Small-scale, or family-size biogas digesters are most frequently found in India and China. However, the demand for such units is growing rapidly throughout the world thanks to more advanced and convenient technologies, such as HomeBiogas. As the modern world is producing more and more waste, individuals are eager to find ecologic ways to treat their trash.

Traditional systems typically found in India and China focus on animal waste. Due to a lack of energy in rural areas combined with a surplus of animal manure, biogas digesters are very popular, useful, and even life-changing. In many developing countries, biogas digesters are even subsidized and advocated by the government and local ministries, who see the variety of benefits produced from using biogas. In addition to having a clean renewable energy provide gas in the kitchen, many families make extensive use of the fertilizer by-product that biogas digesters provide.

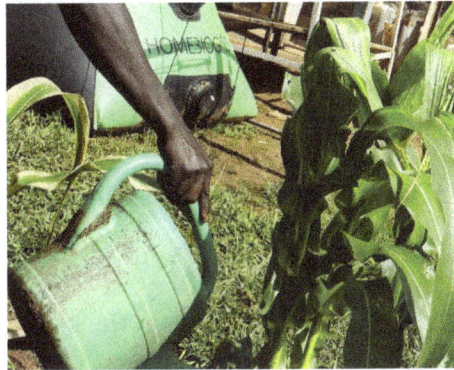

In African countries, some biogas users even turn a profit by selling the bio-slurry by-product produced by biogas systems. This bio-slurry is different from the liquid fertilizer that is produced daily. Bio-slurry refers to the most decomposed stage of the organic matter, after it has been broken down in the system. Bio-slurry sinks to the bottom of the biogas system, and with the help of modern units like HomeBiogas, is easily emptied out once accrued (usually an annual process). This bio-slurry is in fact a nutrient-dense sludge that provides lots of benefits to soil, and can increase productivity of vegetable gardens.

Biogas is a technology that mimics nature's ability to give back. Both industrial-size and family-size biogas units are becoming incredibly popular and relevant in today's world. As the application and efficiency grows, biogas can make a significant impact on reducing greenhouse gases. As a clean source of energy and a renewable means of treating organic waste, biogas is applicable both in under-developed and industrialized countries.

Biogas Production

Biogas is produced through the processing of various types of organic waste. It is a renewable and environmentally friendly fuel made from 100% local feedstocks that is suitable for a diversity of uses including road vehicle fuel and industrial uses. The circular-economy impact of biogas production is further enhanced by the organic nutrients recovered in the production process.

Biogas can be produced from a vast variety of raw materials (feedstocks). The biggest role in the biogas production process is played by microbes feeding on the biomass.

Digestion carried out by these microorganisms creates methane, which can be used as it is locally or upgraded to biogas equivalent to natural gas quality, enabling the transport of the biogas over longer distances. Material containing organic nutrients is also produced in the process, and this can be utilized for purposes such as agriculture.

Stages in Biogas Production

Biogas is produced using well-established technology in a process involving several stages:

- Biowaste is crushed into smaller pieces and slurrified to prepare it for the anaerobic digestion process. Slurrifying means adding liquid to the biowaste to make it easier to process.

- Microbes need warm conditions, so the biowaste is heated to around 37 °C.

- The actual biogas production takes place through anaerobic digestion in large tanks for about three weeks.

- In the final stage, the gas is purified (upgraded) by removing impurities and carbon dioxide.

After this, the biogas is ready for use by enterprises and consumers, for example in a liquefied form or following injection into the gas pipeline network.

Turning Diverse Range of Materials into Gas

Biogas production starts from the arrival of feedstocks at the biogas plant. A diverse range of solid as well as sludge-like feedstocks can be used.

Materials suitable for biogas production include:

- Biodegradable waste from enterprises and industrial facilities, such as surplus lactose from the production of lactose-free dairy products.

- Spoiled food from shops.

- Biowaste generated by consumers.

- Sludge from wastewater treatment plants.

- Manure and field biomass from agriculture.

The material is typically delivered to the biogas plant's reception pit by lorry or waste management vehicle.

A delivery of solid matter such as biowaste will next undergo crushing to make its consistency as even as possible. At this point, water containing nutrients obtained from a further stage in the production process is also mixed with the feedstock to take the rate of solid matter down to only around one-tenth of the total volume.

This is also when any unwanted non-biodegradable waste, such as packaging plastic of out-of-date food waste from shops, is separated from the mixture. This waste is taken to a waste treatment facility where it is used to generate heat and electricity. Biomass that has passed through slurrification is combined with biomass delivered in the form of slurry to the biogas plant and pumped into the pre-digester tank where enzymes secreted by bacteria break down the biomass into an even finer consistency.

Next, the biomass is sanitized before entering the actual biogas reactor (digester). In sanitization, any harmful bacteria found in the material are eliminated by heating the mixture to above 70 °C for one hour. Once sanitized, the mass is pumped into the main reactor where biogas production takes place. Sanitization makes it possible to use the fertilizer product in agriculture.

Biomass is Turned into Gas by Microbes

In the biogas reactor, microbial action begins and the biomass enters a gradual process of fermentation.

In practice this means that microbes feed on the organic matter, such as proteins, carbohydrates and lipids, and their digestion turns these into methane and carbon dioxide.

Most of the organic matter is broken down into biogas – a mixture of methane and carbon dioxide – in approximately three weeks. The biogas is collected in a spherical gas holder from the top of the biogas reactors.

Digestate utilized as Fertilizers or Gardening Soil

The residual solids and liquids created in biogas production are referred to as digestate. This digestate goes into a post-digester reactor and from there further into storage tanks. Digestates are well suited for uses such as fertilization of fields.

Digestates can also be centrifuged to separate the solid and liquid parts.

Solid digestates have uses such as fertilizers in agriculture or in landscaping and can also be turned into gardening soil through a process of maturation involving composting.

Digestates are centrifuged to yield enough process water for the slurrification of biowaste at the beginning of the process. This helps reduce the use of clean water. The centrifuged liquid is rich in nutrients, particularly nitrogen, that can be separated further using methods such as stripping technology and used as fertilizers or nutrient sources in industrial processes.

Clean Biogas helps move Towards Low-carbon Society

Gas would already be ready for several uses straight from the biogas plant gas holder. However, before being injected into the gas pipeline network or used to fuel vehicles, it will still undergo purification.

In this upgrading process, gas is filtered and flown into columns where it is scrubbed by cascading water at a very specific pressure and temperature. Water efficiently absorbs carbon dioxide and sulfur compounds contained by the gas.

Biogas can also be purified using other methods, such as passing it through activated carbon filters to remove impurities.

The final upgraded biogas injected into the gas network is at least 95% and usually around 98% methane. Upgraded biogas still contains a couple of per cent of carbon dioxide as its further separation from methane is not cost-effective let alone sensible as regards the usability of the gas. Biogas is dried carefully before injection into the gas network to prevent condensation in winter subzero conditions.

The biogas produced can be used for purposes such as fuelling municipal waste management vehicles, urban buses or private cars. At the same time, gas serves as evidence of those practical actions that are taking us towards the low-carbon society of the future.

Syngas

Syngas is an abbreviation for synthesis gas, which is a mixture comprising of carbon monoxide, carbon dioxide, and hydrogen. The syngas is produced by gasification of a carbon containing fuel to a gaseous product that has some heating value. Some of the examples of syngas production include gasification of coal emissions, waste emissions to energy gasification, and steam reforming of coke.

The name syngas is derived from the use as an intermediate in generating synthetic natural gas and to create ammonia or methanol. It is a gas that can be used to synthesize other chemicals, hence the name synthesis gas, which was shortened to syngas. Syngas is also an intermediate in creating synthetic petroleum to use as a lubricant or fuel.

Syngas has 50% of the energy density of natural gas. It cannot be burnt directly, but is used as a fuel source. The other use is as an intermediate to produce other chemicals. The production of syngas for use as a raw material in fuel production is accomplished by the gasification of coal or municipal waste. In these reactions, carbon combines with water or oxygen to give rise to carbon dioxide, carbon monoxide, and hydrogen. Syngas is used as an intermediate in the industrial synthesis of ammonia and fertilizer. During this process, methane (from natural gas) combines with water to generate carbon monoxide and hydrogen.

The gasification process is used to convert any material that has carbon to longer hydrocarbon chains. One of the uses of this syngas is as a fuel to manufacture steam or electricity. Another use is as a basic chemical building block for many petrochemical and refining processes.

The general raw materials used for gasification (creation of syngas) are coal, petroleum based materials, or other materials that would be rejected as waste. From these materials, a feedstock is prepared. This is inserted to the gasifier in dry or slurry form. In the gasifier, this feedstock reacts in an oxygen starved environment with steam at elevated pressure and temperature. The resultant syngas is composed of 85% carbon monoxide and hydrogen and small amounts of methane and carbon dioxide.

The syngas may contain some trace elements of impurities, which are removed through further processing and either recovered or redirected to the gasifier. For example, sulfur is recovered in the elemental form or as sulfuric acid and both of these can be marketed. Syngas is a primary source of sulfuric acid. If syngas contains a considerable quantity of nitrogen, the nitrogen must be separated to avoid production of nitric oxides, which are pollutants and contribute to acid rain production. Both carbon monoxide and nitrogen have similar boiling points so recovering pure carbon monoxide requires cryogenic processing, which is very difficult.

If the syngas is to be put to use to generate electricity, then it is generally used as a fuel in an IGCC (integrated gasification combine cycle) power generation configuration. The energy is then utilized by the factor that original produce the syngas, thereby lowering operating costs. There are commercially available technologies to process syngas to generate industrial gases, fertilizers, chemicals, fuels and other products.

Syngas is produced as a result of gasification of a carbon-containing fuel to a gaseous product that has heating value. If syngas contains nitrogen, it must be separated, as both nitrogen and carbon monoxide have similar boiling points and it will be difficult to recover pure carbon monoxide through cryogenic processing.

Production of Syngas

The production of syngas includes the following phases:

The Heating Phase

The first step is gasification, a thermo-chemical process in which carbon-rich feedstocks like petro-coke, biomass or coal are converted into a gaseous compound consisting of carbon monoxide and hydrogen under high-heat, high pressure, oxygen depleted conditions.

Very high temperatures of gasification, normally between 800 and 1500 °C (1472 and 2732 °F) are achieved with the help of an external heat source or through partial oxidation of feedstock which releases heat.

The Reaction Phase

The feedstock reacts with carbon dioxide, water vapor and oxygen during gasification. The reaction is triggered by thermal decomposition for oxygen-rich materials.

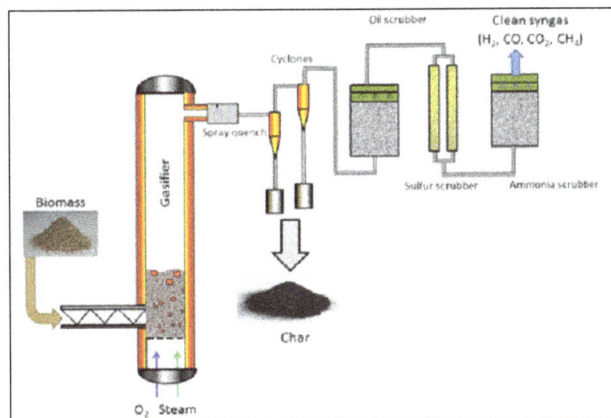

The process flow for Syngas production by gasification of biomass.

The Purificatio Phase

The gas obtained from gasification is raw and not clean enough to use. A purification process is carried out to eliminate impurities like ash, tar, sulfur compounds, methane, water vapor and carbon dioxide. The hydrogen-oxygen proportion is adjusted after purification depending on the applications of the synthesis processes.

The Catalytic Phase

Metals such as iron, manganese, cobalt, copper and new complex molecules are formed when the syngas is in contact with different catalysts. Scientists are experimenting with several catalysts to find new ways of creating already existing molecular combinations. In this way, it is possible to create eco-friendly fuels from syngas.

Syngas Cleanup and Conditioning

Raw syngas obtained from the gasification process needs to be cleaned to eliminate the contaminants such as mercury, chlorides, ammonia, sulfur, fine particulates and other trace heavy metals to protect downstream processes and to meet environmental emission regulations.

Syngas may be conditioned to adjust the hydrogen-to-carbon monoxide ratio based on the downstream process application.

Typical syngas cleanup and conditioning processes include the following:

- Removing bulk particulates using cyclone and filters.

- Wet scrubbing for eliminating chlorides, ammonia and fine particulates.

- Removing trace heavy metal and mercury using solid absorbents.

- Water gas shift for adjusting hydrogen-to-carbon monoxide ratio.

- Catalytic hydrolysis for converting carbonyl sulfide to hydrogen sulfide.

- Acid gas removal for extracting sulfur-bearing gases and carbon dioxide.

Syngas Fermentation

Syngas fermentation is a microbial process in which syngas is used as a carbon and energy source, and then converted into chemicals and fuels with the help of microorganisms. Methane, butyric acid, acetic acid, butanol and ethanol are the main products of syngas fermentation.

Acetogens such as Clostridium carboxidivorans, Eurobacterium limosum, Butyribacterium methylotrophicum and Peptostreptococcus products are involved in the production of chemicals and fuels.

Some of the key benefits of syngas fermentation process include the following:

- High reaction specificity.

- Low temperature and pressure.

- Does not require a specific ratio of CO to H_2.

- Tolerate compounds having high sulfur content.

However, syngas fermentation has certain limitations such as inhibition of organisms, low volumetric productivity and gas-liquid mass transfer limitation.

Applications

Syngas can be used to produce a wide range of fertilizers, fuels, solvent and synthetic materials. Few examples are as follows:

- Steam for use in turbine drivers for electricity generation.

- Nitrogen for use as pressurizing agents and fertilizers.

- Hydrogen for electricity generation, use in refinery industry to extract more diesel and gasoline from crude oil and for a large variety of hydrogenation reactions where hydrogen is added to unsaturated hydrocarbons.

- Ammonia for use as fertilizers and for the production of plastics like polyurethane and nylon.

- Methanol for the production of plastics, resins, pharmaceuticals, adhesives, paints and also as a component of fuels.

- Carbon monoxide for use in chemical industry feedstock and fuels.

- Sulfur for use as elemental sulfur for chemical industry.

- Minerals and solids for use as slag for roadbeds.

Synthetic Fuel

Synthetic fuel or synfuel is a liquid fuel, or sometimes gaseous fuel, obtained from syngas, a mixture of carbon monoxide and hydrogen, in which the syngas was derived from gasification of solid feedstocks such as coal or biomass or by reforming of natural gas.

Common methods for refining synthetic fuels include the Fischer–Tropsch conversion, methanol to gasoline conversion, or direct coal liquefaction.

As of July 2009, worldwide commercial synthetic fuels production capacity was over 240,000 barrels per day (38,000 m^3/d), with numerous new projects in construction or development.

Classificatio and Principles

The term 'synthetic fuel' or 'synfuel' has several different meanings and it may include different types of fuels. More traditional definitions, such as the definition given by the International Energy Agency, define 'synthetic fuel' or synfuel' as any liquid fuel obtained from coal or natural gas. In its Annual Energy Outlook 2006, the Energy Information Administration defines synthetic fuels as fuels produced from coal, natural gas, or biomass feedstocks through chemical conversion into synthetic crude and/or synthetic liquid products. A number of synthetic fuel's definitions include fuels produced from biomass, and industrial and municipal waste. The definition of synthetic fuel also allows oil sands and oil shale as synthetic fuel sources, and in addition to liquid fuels, synthesized gaseous fuels are also considered to be synthetic fuels: in his 'Synthetic fuels handbook' petrochemist James G. Speight included liquid and gaseous fuels as well as clean solid fuels produced by conversion of coal, oil shale or tar sands, and various forms of biomass, although he admits that in the context of substitutes for petroleum-based fuels it has even wider meaning. Depending on the context, methanol, ethanol and hydrogen may also be included.

Synthetic fuels are produced by the chemical process of conversion. Conversion methods could be direct conversion into liquid transportation fuels, or indirect conversion, in which the source substance is converted initially into syngas which then goes through additional conversion process to become liquid fuels. Basic conversion methods include carbonization and pyrolysis, hydrogenation, and thermal dissolution.

Processes

The numerous processes that can be used to produce synthetic fuels broadly fall into three categories: Indirect, Direct, and Biofuel processes.

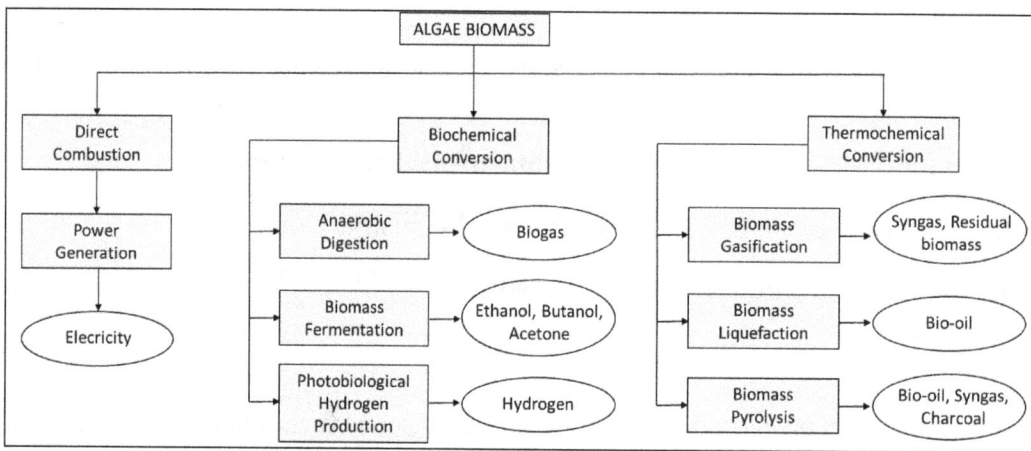

Indirect Conversion

Indirect conversion has the widest deployment worldwide, with global production totaling around 260,000 barrels per day (41,000 m³/d), and many additional projects under active development.

Indirect conversion broadly refers to a process in which biomass, coal, or natural gas is converted to a mix of hydrogen and carbon monoxide known as syngas either through gasification or steam methane reforming, and that syngas is processed into a liquid transportation fuel using one of a number of different conversion techniques depending on the desired end product.

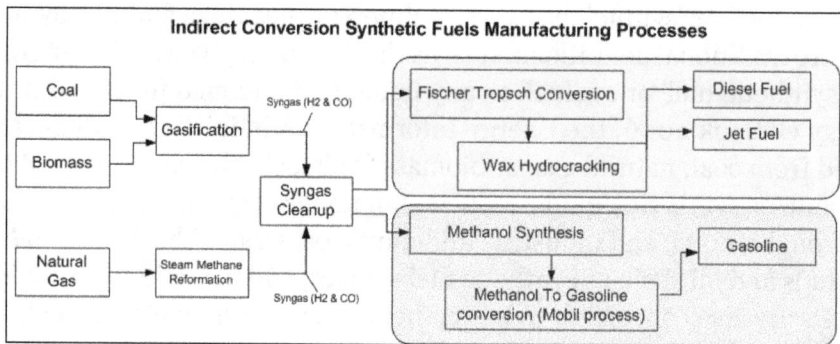

The primary technologies that produce synthetic fuel from syngas are Fischer–Tropsch synthesis and the Mobil process (also known as Methanol-To-Gasoline, or MTG). In the Fischer–Tropsch process syngas reacts in the presence of a catalyst, transforming into liquid products (primarily diesel fuel and jet fuel) and potentially waxes (depending on the FT process employed).

The process of producing synfuels through indirect conversion is often referred to as coal-to-liquids (CTL), gas-to-liquids (GTL) or biomass-to-liquids (BTL), depending on the initial feedstock. At least three projects (Ohio River Clean Fuels, Illinois Clean Fuels, and Rentech Natchez) are combining coal and biomass feedstocks, creating hybrid-feedstock synthetic fuels known as Coal and Biomass To Liquids (CBTL).

Indirect conversion process technologies can also be used to produce hydrogen, potentially for use in fuel cell vehicles, either as slipstream co-product, or as a primary output.

Direct Conversion

Direct conversion refers to processes in which coal or biomass feedstocks are converted directly into intermediate or final products, avoiding the conversion to syngas via gasification. Direct conversion processes can be broadly broken up into two different methods: Pyrolysis and carbonization, and hydrogenation.

Hydrogenation Processes

One of the main methods of direct conversion of coal to liquids by hydrogenation process is the Bergius process. In this process, coal is liquefied by heating in the presence of hydrogen gas (hydrogenation). Dry coal is mixed with heavy oil recycled from the process. Catalysts are typically added to the mixture. The reaction occurs at between 400 °C (752 °F) to 500 °C (932 °F) and 20 to 70 MPa hydrogen pressure. The reaction can be summarized as follows:

$$n\mathrm{C} + (n+1)\mathrm{H}_2 \rightarrow \mathrm{C}_n\mathrm{H}_{2n+2}$$

After World War I several plants were built in Germany; these plants were extensively used during World War II to supply Germany with fuel and lubricants.

The Kohleoel Process, developed in Germany by Ruhrkohle and VEBA, was used in the demonstration plant with the capacity of 200 ton of lignite per day, built in Bottrop, Germany. This plant operated from 1981 to 1987. In this process, coal is mixed with a recycle solvent and iron catalyst. After preheating and pressurizing, H_2 is added. The process takes place in tubular reactor at the pressure of 300 bar and at the temperature of 470 °C (880 °F). This process was also explored by SASOL in South Africa.

In 1970-1980s, Japanese companies Nippon Kokan, Sumitomo Metal Industries and Mitsubishi Heavy Industries developed the NEDOL process. In this process, a mixture of coal and recycled solvent is heated in the presence of iron-based catalyst and H_2. The reaction takes place in tubular reactor at temperature between 430 °C (810 °F) and 465 °C (870 °F) at the pressure 150-200 bar. The produced oil has low quality and requires intensive upgrading. H-Coal process, developed by Hydrocarbon Research, Inc., in 1963, mixes pulverized coal with recycled liquids, hydrogen and catalyst in the ebullated bed reactor. Advantages of this process are that dissolution and oil upgrading are taking place in the single reactor, products have high H:C ratio, and a fast reaction time, while the main disadvantages are high gas yield, high hydrogen consumption, and limitation of oil usage only as a boiler oil because of impurities.

The SRC-I and SRC-II (Solvent Refined Coal) processes were developed by Gulf Oil and implemented as pilot plants in the United States in the 1960s and 1970s. The Nuclear Utility Services Corporation developed hydrogenation process which was patented by Wilburn C. Schroeder in 1976. The process involved dried, pulverized coal mixed with roughly 1wt% molybdenum catalysts. Hydrogenation occurred by use of high temperature and pressure syngas produced in a separate gasifier. The process ultimately yielded a synthetic crude product, Naphtha, a limited amount of C_3/C_4 gas, light-medium weight liquids (C_5-C_{10}) suitable for use as fuels, small amounts of NH_3 and significant amounts of CO_2. Other single-stage hydrogenation processes are the Exxon donor solvent process, the Imhausen High-pressure Process, and the Conoco Zinc Chloride Process.

A number of two-stage direct liquefaction processes have been developed. After the 1980s only the Catalytic Two-stage Liquefaction Process, modified from the H-Coal Process; the Liquid Solvent Extraction Process by British Coal; and the Brown Coal Liquefaction Process of Japan have been developed.

Chevron Corporation developed a process invented by Joel W. Rosenthal called the Chevron Coal Liquefaction Process (CCLP). It is unique due to the close-coupling of the non-catalytic dissolver and the catalytic hydroprocessing unit. The oil produced had properties that were unique when compared to other coal oils; it was lighter and had far fewer heteroatom impurities. The process was scaled-up to the 6 ton per day level, but not proven commercially.

Pyrolysis and Carbonization Processes

There are a number of different carbonization processes. The carbonization conversion occurs through pyrolysis or destructive distillation, and it produces condensable coal tar, oil and water vapor, non-condensable synthetic gas, and a solid residue-char. The condensed coal tar and oil are then further processed by hydrogenation to remove sulfur and nitrogen species, after which they are processed into fuels.

The typical example of carbonization is the Karrick process. The process was invented by Lewis Cass Karrick in the 1920s. The Karrick process is a low-temperature carbonization process, where coal is heated at 680 °F (360 °C) to 1,380 °F (750 °C) in the absence of air. These temperatures optimize the production of coal tars richer in lighter hydrocarbons than normal coal tar. However, the produced liquids are mostly a by-product and the main product is semi-coke, a solid and smokeless fuel.

The COED Process, developed by FMC Corporation, uses a fluidized bed for processing, in combination with increasing temperature, through four stages of pyrolysis. Heat is transferred by hot gases produced by combustion of part of the produced char. A modification of this process, the COGAS Process, involves the addition of gasification of char. The TOSCOAL Process, an analogue to the TOSCO II oil shale retorting process and Lurgi-Ruhrgas process, which is also used for the shale oil extraction, uses hot recycled solids for the heat transfer.

Liquid yields of pyrolysis and Karrick processes are generally low for practical use for synthetic liquid fuel production. Furthermore, the resulting liquids are of low quality and require further treatment before they can be used as motor fuels. In summary, there is little possibility that this process will yield economically viable volumes of liquid fuel.

Biofuels Processes

One example of a Biofuel-based synthetic fuel process is Hydrotreated Renewable Jet (HRJ) fuel. There are a number of variants of these processes under development, and the testing and certification process for HRJ aviation fuels is beginning.

There are two such process under development by UOP. One using solid biomass feedstocks, and one using bio-oil and fats. The process using solid second-generation biomass sources such as switchgrass or woody biomass uses pyrolysis to produce a bio-oil, which is then catalytically stabilized and deoxygenated to produce a jet-range fuel. The process using natural oils and fats goes through a deoxygenation process, followed by hydrocracking and isomerization to produce a renewable Synthetic Paraffinic Kerosene jet fuel.

Oil Sand and Oil Shale Processes

Synthetic crude may also be created by upgrading bitumen (a tar like substance found in oil sands), or synthesizing liquid hydrocarbons from oil shale. There are a number of processes extracting shale oil (synthetic crude oil) from oil shale by pyrolysis, hydrogenation, or thermal dissolution.

Commercialization

Worldwide commercial synthetic fuels plant capacity is over 240,000 barrels per day (38,000 m^3/d), including indirect conversion Fischer–Tropsch plants in South Africa (Mossgas, Secunda CTL), Qatar {Oryx GTL}, and Malaysia (Shell Bintulu), and a Mobil process (Methanol to Gasoline) plant in New Zealand.

Sasol, a company based in South Africa operates the world's only commercial Fischer–Tropsch coal-to-liquids facility at Secunda, with a capacity of 150,000 barrels per day (24,000 m^3/d).

Economics

The economics of synthetic fuel manufacture vary greatly depending the feedstock used, the precise process employed, site characteristics such as feedstock and transportation costs, and the cost of additional equipment required to control emissions. The examples described below indicate a wide range of production costs between $20/BBL for large-scale gas-to-liquids, to as much as $240/BBL for small-scale biomass-to-liquids + Carbon Capture and Sequestration.

In order to be economically viable, projects must do much better than just being competitive head-to-head with oil. They must also generate a sufficient return on investment to justify the capital investment in the project.

CTL/CBTL/BTL Economics

According to a December 2007 study, a medium scale (30,000 BPD) coal-to-liquids plant (CTL) sited in the US using bituminous coal, is expected to be competitive with oil down to roughly $52–56/bbl crude-oil equivalent. Adding carbon capture and sequestration to the project was expected to add an additional $10/BBL to the required selling price, though this may be offset by revenues from enhanced oil recovery, or by tax credits, or the eventual sale of carbon credits.

A recent NETL study examined the relative economics of a number of different process configurations for the production of indirect FT fuels using biomass, coal, and CCS. This study determined a price at which the plant would not only be profitable, but also make a sufficient return to yield a 20% return on the equity investment required to build the plant.

This chapter details an analysis which derives the Required Selling Price (RSP) of the FT diesel fuels produced in order to determine the economic feasibility and relative competitiveness of the different plant options. A sensitivity analysis was performed to determine how carbon control regulations such as an emissions trading scheme for transportation fuels would affect the price of both petroleum-derived diesel and FT diesel from the different plants. The key findings of these analyses were: (1) CTL plants equipped with CCS are competitive at crude oil prices as low as $86 per barrel and have less life cycle GHG emissions than petroleum-derived diesel. These plants become more economically competitive as carbon prices increase. (2) The incremental cost of adding simple CCS is very low (7 cents per gallon) because CO_2 capture is an inherent part of the FT process. This becomes the economically preferred option at carbon prices above $5/mtCO_2$. (3) BTL systems are hindered by limited biomass availability which affects the maximum plant size, thereby limiting potential economies of scale. This, combined with relatively high biomass costs results in FT diesel prices which are double that of other configurations: $6.45 to $6.96/gal compared to $2.56 to $2.82/gal for CTL and 15wt% CBTL systems equipped with CCS.

The conclusion reached based on these findings was that both the CTL with CCS and the 8wt% to 15wt% CBTL with CCS configurations may offer the most pragmatic solutions to the nation's energy strategy dilemma: GHG emission reductions which are significant (5% to 33% below the petroleum baseline) at diesel RSPs that are only half as much as the BTL options ($2.56 to $2.82 per gallon compared to $6.45 to $6.96 per gallon for BTL). These options are economically feasible when crude oil prices are $86 to $95 per barrel.

These economics can change in the event that plentiful low-cost biomass sources can be found, lowing the cost of biomass inputs, and improving economies of scale.

Economics for solid feedstock indirect FT process plants are further confused by carbon regulation. Generally, since permitting a CTL plant without CCS will likely be impossible, and CTL+CCS plants have a lower carbon footprint than conventional fuels, carbon regulation is expected to be balance-positive for synthetic fuel production. But it impacts the economics of different process configurations in different ways. The NETL study picked a blended CBTL process using 5-15% biomass alongside coal as the most economical in a range of carbon price and probable future regulation scenarios. Unfortunately, because of scale and cost constraints, pure BTL processes did not score well until very high carbon prices were assumed, though again this may improve with better feedstocks and more efficient larger scale projects.

Chinese Direct Coal Liquefaction Economics

Shenhua Group recently reported that their direct coal liquefaction process is competitive with oil prices above $60 per barrel.< Previous reports have indicated an anticipated cost of production of less than $30 per barrel, based on a direct coal liquefaction process, and a coal mining cost of under $10/ton. In October 2011, actual price of coal in China was as high as $135/ton.

Security Considerations

A central consideration for the development of synthetic fuel is the security factor of securing domestic fuel supply from domestic biomass and coal. Nations that are rich in biomass and coal can use synthetic fuel to off-set their use of petroleum derived fuels and foreign oil.

Environmental Considerations

The environmental footprint of a given synthetic fuel varies greatly depending on which process is employed, what feedstock is used, what pollution controls are employed, and what the transportation distance and method are for both feedstock procurement and end-product distribution.

In many locations, project development will not be possible due to permitting restrictions if a process design is chosen that does not meet local requirements for clean air, water, and increasingly, lifecycle carbon emissions.

Lifecycle Greenhouse Gas Emissions

Among different indirect FT synthetic fuels production technologies, potential emissions of greenhouse gasses vary greatly. Coal to liquids ("CTL") without carbon capture and sequestration ("CCS") is expected to result in a significantly higher carbon footprint than conventional petroleum-derived fuels (+147%). On the other hand, biomass-to-liquids with CCS could deliver a 358% reduction in lifecycle greenhouse gas emissions. Both of these plants fundamentally use gasification and FT conversion synthetic fuels technology, but they deliver wildly divergent environmental footprints.

Lifecycle carbon emissions profiles of various fuels, including many synthetic fuels. Coal and biomass co-conversion to transportation fuels, Michael E. Reed, DOE NETL Office of Fossil Energy, Oct 17 2007

Generally, CTL without CCS has a higher greenhouse gas footprint. CTL with CCS has a 9-15% reduction in lifecycle greenhouse gas emissions compared to that of petroleum derived diesel.

CBTL+CCS plants that blend biomass alongside coal while sequestering carbon do progressively better the more biomass is added. Depending on the type of biomass, the assumptions about root storage, and the transportation logistics, at conservatively 40% biomass alongside coal, CBTL+CCS plants achieve a neutral lifecycle greenhouse gas footprint. At more than 40% biomass, they begin to go lifecycle negative, and effectively store carbon in the ground for every gallon of fuels that they produce.

Ultimately BTL plants employing CCS could store massive amounts of carbon while producing transportation fuels from sustainably produced biomass feedstocks, although there are a number of significant economic hurdles, and a few technical hurdles that would have to be overcome to enable the development of such facilities.

Serious consideration must also be given to the type and method of feedstock procurement for either the coal or biomass used in such facilities, as reckless development could exacerbate environmental problems caused by mountaintop removal mining, land use change, fertilizer runoff, food vs. fuels concerns, or many other potential factors. Or they could not, depending entirely on project-specific factors on a plant-by-plant basis.

A study from U.S. Department of Energy National Energy Technology Laboratory with much more in-depth information of CBTL life-cycle emissions "Affordable Low Carbon Diesel from Domestic Coal and Biomass".

Hybrid hydrogen-carbon processes have also been proposed recently as another closed-carbon cycle alternative, combining 'clean' electricity, recycled CO, H_2 and captured CO_2 with biomass as inputs as a way of reducing the biomass needed.

Fuels Emissions

The fuels produced by the various synthetic fuels process also have a wide range of potential environmental performance, though they tend to be very uniform based on the type of synthetic fuels process used (i.e. the tailpipe emissions characteristics of Fischer–Tropsch diesel tend to be the same, though their lifecycle greenhouse gas footprint can vary substantially based on which plant produced the fuel, depending on feedstock and plant level sequestration considerations).

In particular, Fischer–Tropsch diesel and jet fuels deliver dramatic across-the-board reductions in all major criteria pollutants such as SOx, NOx, Particulate Matter, and Hydrocarbon emissions. These fuels, because of their high level of purity and lack of contaminants, further enable the use of advanced emissions control equipment that has been shown to virtually eliminate HC, CO, and PM emissions from diesel vehicles.

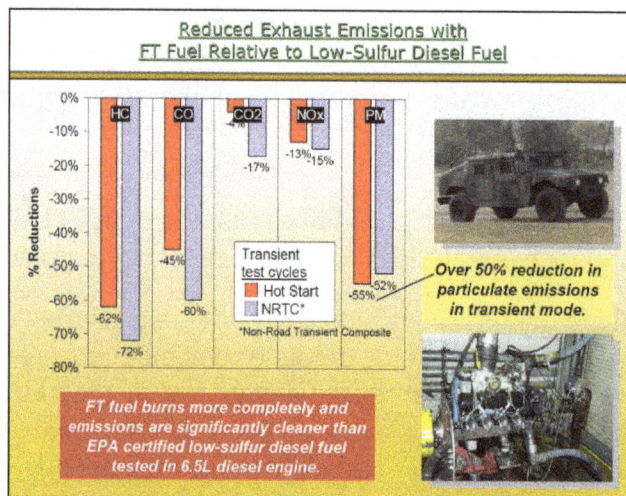

Using Fischer–Tropsch diesel results in dramatic across the board tailpipe
emissions reductions relative to conventional fuels.

In testimony before the Subcommittee on Energy and Environment of the U.S. House of Representatives the following statement was made by a senior scientist from Rentech:

"F-T fuels offer numerous benefits to aviation users. The first is an immediate reduction in particulate emissions. F-T jet fuel has been shown in laboratory combusters and engines to reduce PM emissions by 96% at idle and 78% under cruise operation. Validation of the reduction in other turbine engine emissions is still under way. Concurrent to the PM reductions is an immediate reduction in CO_2 emissions from F-T fuel. F-T fuels inherently reduce CO_2 emissions because they have higher energy content per carbon content of the fuel, and the fuel is less dense than conventional jet fuel allowing aircraft to fly further on the same load of fuel."

The "cleanness" of these FT synthetic fuels is further demonstrated by the fact that they are sufficiently non-toxic and environmentally benign as to be considered biodegradable. This owes primarily to the near-absence of sulfur and extremely low level of aromatics present in the fuel.

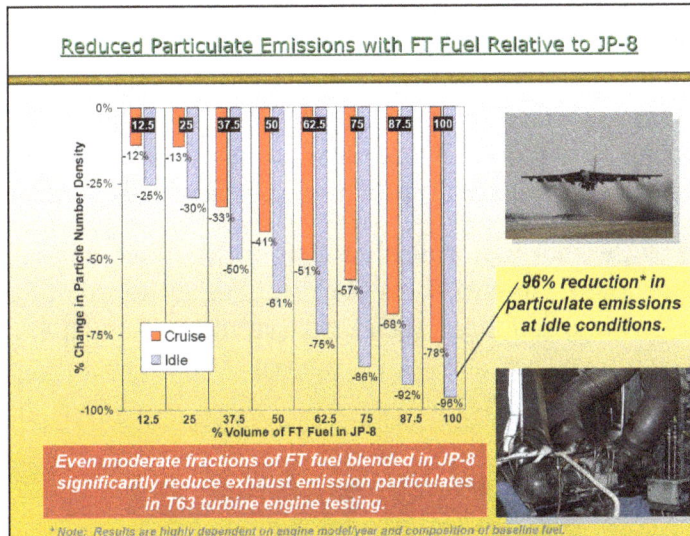

Using Fischer–Tropsch jet fuels have been proven to dramatically reduce particulate and other aircraft emissions.

Sustainability

One concern commonly raised about the development of synthetic fuels plants is sustainability. Fundamentally, transitioning from oil to coal or natural gas for transportation fuels production is a transition from one inherently depletable geologically limited resource to another.

One of the positive defining characteristics of synthetic fuels production is the ability to use multiple feedstocks (coal, gas, or biomass) to produce the same product from the same plant. In the case of hybrid BCTL plants, some facilities are already planning to use a significant biomass component alongside coal. Ultimately, given the right location with good biomass availability, and sufficiently high oil prices, synthetic fuels plants can be transitioned from coal or gas, over to a 100% biomass feedstock. This provides a path forward towards a renewable fuel source and possibly more sustainable, even if the plant originally produced fuels solely from coal, making the infrastructure forwards-compatible even if the original fossil feedstock runs out.

Some synthetic fuels processes can be converted to sustainable production practices more easily than others, depending on the process equipment selected. This is an important design consideration as these facilities are planned and implemented, as additional room must be left in the plant layout to accommodate whatever future plant change requirements in terms of materials handling and gasification might be necessary to accommodate a future change in production profile.

Alcohol Fuel

Alcohols have been used as a fuel. The first four aliphatic alcohols (methanol, ethanol, propanol, and butanol) are of interest as fuels because they can be synthesized chemically or biologically, and they have characteristics which allow them to be used in internal combustion engines. The general chemical formula for alcohol fuel is $C_nH_{2n+1}OH$.

Most methanol is produced from natural gas, although it can be produced from biomass using very similar chemical processes. Ethanol is commonly produced from biological material through fermentation processes. Biobutanol has the advantage in combustion engines in that its energy density is closer to gasoline than the simpler alcohols (while still retaining over 25% higher octane rating); however, biobutanol is currently more difficult to produce than ethanol or methanol. When obtained from biological materials and/or biological processes, they are known as bioalcohols (e.g. "bioethanol"). There is no chemical difference between biologically produced and chemically produced alcohols.

One advantage shared by the four major alcohol fuels is their high octane rating. This tends to increase their fuel efficiency and largely offsets the lower energy density of vehicular alcohol fuels (as compared to petrol/gasoline and diesel fuels), thus resulting in comparable "fuel economy" in terms of distance per volume metrics, such as kilometers per liter, or miles per gallon.

Methanol and Ethanol

Ethanol used as a fuel.

Methanol and ethanol can both be derived from fossil fuels, biomass, or perhaps most simply, from carbon dioxide and water. Ethanol has most commonly been produced through fermentation of sugars, and methanol has most commonly been produced from synthesis gas, but there are more modern ways to obtain these fuels. Enzymes can be used instead of fermentation. Methanol is the simpler molecule, and ethanol can be made from methanol. Methanol can be produced industrially from nearly any biomass, including animal waste, or from carbon dioxide and water or steam by first converting the biomass to synthesis gas in a gasifier. It can also be produced in a laboratory using electrolysis or enzymes.

As a fuel, methanol and ethanol both have advantages and disadvantages over fuels such as petrol (gasoline) and diesel fuel. In spark ignition engines, both alcohols can run at a much higher exhaust gas recirculation rates and with higher compression ratios. Both alcohols have a high octane rating, with ethanol at 109 RON (Research Octane Number), 90 MON (Motor Octane Number), (which equates to 99.5 AKI) and methanol at 109 RON, 89 MON (which equates to 99 AKI). Note that AKI refers to 'Anti-Knock Index' which averages the RON and MON ratings (RON+MON)/2, and is used on U.S. gas station pumps. Ordinary European petrol is typically 95 RON, 85 MON, equal to 90 AKI. As a compression ignition engine fuel, both alcohols create very little particulates, but their low cetane number means that an ignition improver like glycol must be mixed into the fuel with approx. 5%.

When used in spark ignition engines alcohols have the potential to reduce NOx, CO, HC and particulates. A test with E85 fueled Chevrolet Luminas showed that NMHC went down by 20-22%, NOx by 25-32% and CO by 12-24% compared to reformulated gasoline. Toxic emissions of benzene and 1,3 Butadiene also decreased while aldehyde emissions increased (acetaldehyde in particular).

Tailpipe emissions of CO_2 also decrease due to the lower carbon-to-hydrogen ratio of these alcohols, and the improved engine efficiency.

Methanol and ethanol fuels contain soluble and insoluble contaminants. Halide ions, which are soluble contaminants, such as chloride ions, have a large effect on the corrosivity of alcohol fuels. Halide ions increase corrosion in two ways: they chemically attack passivating oxide films on several metals causing pitting corrosion, and they increase the conductivity of the fuel. Increased electrical conductivity promotes electrical, galvanic and ordinary corrosion in the fuel system. Soluble contaminants such as aluminum hydroxide, itself a product of corrosion by halide ions, clogs the fuel system over time.

To prevent corrosion the fuel system must be made of suitable materials, electrical wires must be properly insulated and the fuel level sensor must be of pulse and hold type, magneto resistive or other similar non-contact type. In addition, high quality alcohol should have a low concentration of contaminants and have a suitable corrosion inhibitor added. Scientific evidence reveals that also water is an inhibitor for corrosion by ethanol.

The experiments are done with E50, which is more aggressive & speeds up the corrosion effect. It is very clear that by increasing the amount of water in fuel ethanol one can reduce the corrosion. At 2% or 20,000 ppm water in the fuel ethanol the corrosion stopped. The observations in Japan are in line with the fact that hydrous ethanol is known for being less corrosive than anhydrous ethanol. The reaction mechanism is 3 EtOH + Al -> $Al(OEt)_3$ + ³⁄₂ H_2 will be the same at lower-mid blends. When enough water is present in the fuel, the aluminum will react preferably with water to produce Al_2O_3, repairing the protective aluminum oxide layer. The aluminum alkoxide does not make a tight oxide layer; water is essential to repair the holes in the oxide layer.

Methanol and ethanol are also incompatible with some polymers. The alcohol reacts with the polymers causing swelling, and over time the oxygen breaks down the carbon-carbon bonds in the polymer causing a reduction in tensile strength. For the past few decades though, most cars have been designed to tolerate up to 10% ethanol (E10) without problem. This includes both fuel system compatibility and lambda compensation of fuel delivery with fuel injection engines featuring closed loop lambda control. In some engines ethanol may degrade some compositions of plastic or rubber fuel delivery components designed for conventional petrol, and also be unable to lambda compensate the fuel properly.

"FlexFuel" vehicles have upgraded fuel system and engine components which are designed for long life using E85 or M85, and the ECU can adapt to any fuel blend between gasoline and E85 or M85. Typical upgrades include modifications to: fuel tanks, fuel tank electrical wiring, fuel pumps, fuel filters, fuel lines, filler tubes, fuel level sensors, fuel injectors, seals, fuel rails, fuel pressure regulators, valve seats and inlet valves. "Total Flex" Autos destined for the Brazilian market can use E100 (100% Ethanol).

One liter of ethanol contain 21.1 MJ, a liter of methanol 15.8 MJ and a liter of gasoline approximately 32.6 MJ. In other words, for the same energy content as one liter or one gallon of gasoline, one needs 1.6 liters/gallons of ethanol and 2.1 liters/gallons of methanol. The raw energy-per-volume numbers produce misleading fuel consumption numbers however, because alcohol-fueled engines can be made substantially more energy-efficient. A larger percentage of the energy available in a liter of alcohol fuel can be converted to useful work. This difference in efficiency can partially or totally balance out the energy density difference, depending on the particular engines being compared.

Methanol fuel has been proposed as a future biofuel, often as an alternative to the hydrogen economy. Methanol has a long history as a racing fuel. Early Grand Prix Racing used blended mixtures as well as pure methanol. The use of the fuel was primarily used in North America after the war. However, methanol for racing purposes has largely been based on methanol produced from syngas derived from natural gas and therefore this methanol would not be considered a biofuel. Methanol is a possible biofuel, however when the syngas is derived from biomass.

In theory, methanol can also be produced from carbon dioxide and hydrogen using nuclear power or any renewable energy source, although this is not likely to be economically viable on an industrial scale.Compared to bioethanol, the primary advantage of methanol biofuel is its much greater well-to-wheel efficiency. This is particularly relevant in temperate climates where fertilizers are needed to grow sugar or starch crops to make ethanol, whereas methanol can be produced from lignocellulose (woody) biomass.

Ethanol is already being used extensively as a fuel additive, and the use of ethanol fuel alone or as part of a mix with gasoline is increasing. Compared to methanol its primary advantage is that it is less corrosive and additionally the fuel is non-toxic, although the fuel will produce some toxic exhaust emissions. Since 2007, the Indy Racing League has used ethanol as its exclusive fuel, after 40 years of using methanol. Since September 2007 petrol stations in NSW, Australia are mandated to supply all their petrol with 2% Ethanol content

Butanol and Propanol

Propanol and butanol are considerably less toxic and less volatile than methanol. In particular, butanol has a high flash point of 35 °C, which is a benefit for fire safety, but may be a difficulty for starting engines in cold weather. The concept of flash point is however not directly applicable to engines as the compression of the air in the cylinder means that the temperature is several hundred degrees Celsius before ignition takes place.

The fermentation processes to produce propanol and butanol from cellulose are fairly tricky to execute, and the Weizmann organism (Clostridium acetobutylicum) currently used to perform these conversions produces an extremely unpleasant smell, and this must be taken into consideration when designing and locating a fermentation plant. This organism also dies when the butanol content of whatever it is fermenting rises to 7%. For comparison, yeast dies when the ethanol content of its feedstock hits 14%. Specialized strains can tolerate even greater ethanol concentrations - so-called turbo yeast can withstand up to 16% ethanol. However, if ordinary Saccharomyces yeast can be modified to improve its ethanol resistance, scientists may yet one day produce a strain of the Weizmann organism with a butanol resistance higher than the natural boundary of 7%. This would be useful because butanol has a higher energy density than ethanol, and because waste fibre left over from sugar crops used to make ethanol could be made into butanol, raising the alcohol yield of fuel crops without there being a need for more crops to be planted.

Despite these drawbacks, DuPont and BP have recently announced that they are jointly to build a small scale butanol fuel demonstration plant alongside the large bioethanol plant they are jointly developing with Associated British Foods.

The company Energy Environment International developed a method for producing butanol from biomass, which involves the use of two separate micro-organisms in sequence to minimize production of acetone and ethanol byproducts.

The Swiss company Butalco GmbH uses a special technology to modify yeasts in order to produce butanol instead of ethanol. Yeasts as production organisms for butanol have decisive advantages compared to bacteria.

Butanol combustion is: $C_4H_9OH + 6O_2 \rightarrow 4CO_2 + 5H_2O + heat$

Propanol combustion is: $2C_3H_7OH + 9O_2 \rightarrow 6 CO_2 + 8H_2O + heat$

The 3-carbon alcohol, propanol (C_3H_7OH), is not often used as a direct fuel source for petrol engines(unlike ethanol, methanol and butanol), with most being directed into use as a solvent. However, it is used as a source of hydrogen in some types of fuel cell; it can generate a higher voltage than methanol, which is the fuel of choice for most alcohol-based fuel cells. However, since propanol is harder to produce than methanol (biologically OR from oil), methanol-utilising fuel cells are preferred over those that utilise propanol.

Alcohol Fuel in different Parts of the World

Brazil

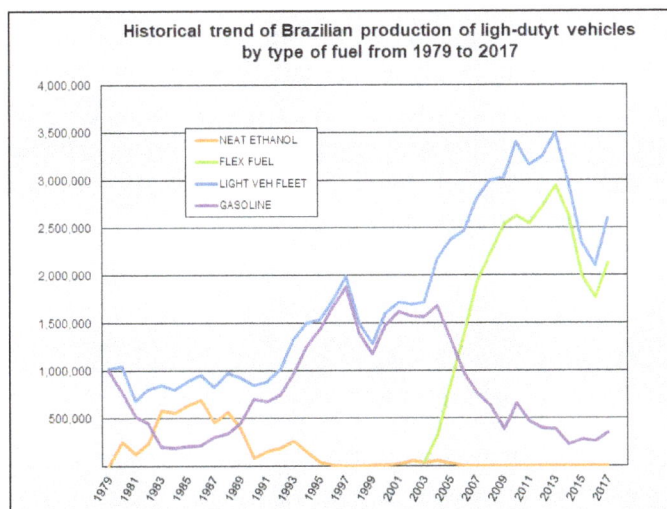

Historical trend of Brazilian production of light vehicles by type of fuel, neat ethanol (alcohol), flex fuel, and gasoline vehicles from 1979 to 2017.

Brazil was until recently the largest producer of alcohol fuel in the world, typically fermenting ethanol from sugarcane.

The country produces a total of 18 billion litres (4.8 billion gallons) annually, of which 3.5 billion liters are exported, 2 billion of them to the U.S.. Alcohol cars debuted in the Brazilian market in 1979 and became quite popular because of heavy subsidy, but in the 1980s prices rose and gasoline regained the leading market share.

However, from 2003 on, alcohol is rapidly rising its market share once again because of new technologies involving flexible-fuel engines, called "Flex" or "Total Flex" by all major car manufacturers (Volkswagen, General Motors, Fiat, etc.). "Flex" engines work with gasoline, alcohol or any mixture of both fuels. As of May 2009, more than 88% of new vehicles sold in Brazil are flex fuel.

Because of the Brazilian leading production and technology, many countries became very interested in importing alcohol fuel and adopting the "Flex" vehicle concept. On March 7 of 2007, US president George W. Bush visited the city of São Paulo to sign agreements with Brazilian president Luiz Inácio Lula da Silva on importing alcohol and its technology as an alternative fuel.

China

As early as 1935, China has made alcohol fuel powered cars. China has reported with a 70% methanol use to conventional gasoline an independence from crude oil.

National Committee of Planning and Action Coordination for Clean Automobile had listed key technologies related to alcohol/ether fuel and accelerated industrialization into its main agenda. Alcohol fuels had become part of five main alternative fuels: Two of which were alcohols; methanol and ethanol

United States

The United States at the end of 2007 was producing 26.9 billion litres (7 billion gallons) per year. E10 or Gasohol is commonly marketed in Delaware and E85 is found in many states, particularly in the Midwest where ethanol from corn is produced locally.

Many states and municipalities have mandated that all gasoline fuel be blended with 10 percent alcohol (usually ethanol) during some or all of the year. This is to reduce pollution and allows these areas to comply with federal pollution limits. Because alcohol is partially oxygenated, it produces less overall pollution, including ozone. In some areas (California in particular) the regulations may also require other formulations or added chemicals that reduce pollution, but add complexity to the fuel distribution and increase the cost of the fuel.

Japan

The first alcohol fuel in Japan began with GAIAX in 1999. GAIAX was developed in South Korea, and imported by Japan. The principal ingredient was methanol.

Because GAIAX was not gasoline, it was a tax-free object of the gas tax of Japan. However, as a result, the use of GAIAX came to be considered an act of smuggling in Japan by the government and the petroleum industry. Retailing of GAIAX was done to avoid the tax evasion criticism by independently paying the diesel fuel tax in the legal system regulations.

Accidental vehicle fires where GAIAX was being refueled began to be reported in around 2000 when the tax evasion discussion had almost ended. The car industry in Japan criticized GAIAX, saying that "fires broke out because high density alcohol had corroded the fuel pipes". GAIAX was named a "high density alcohol fuel," and a campaign was executed to exclude it from the market long term. Finally, the Ministry of Economy, Trade and Industry also joined this campaign.

The gasoline quality method was revised under the pretext of safety concerns in 2003. This prohibited the manufacturing and sale of "High density alcohol fuel", and added a substantial GAIAX sales ban. By revising the law, fuel manufacturers are prohibited from adding 3% or more alcohol to gasoline. This revision to the law is grounds not to be able to sell alcohol fuel greater than E3 in Japan.

The petroleum industry in Japan is now proceeding with research and development of an original alcohol fuel that differs from GAIAX. However, the commercial manufacture and sale of any new fuel may be barred by existing laws that currently exclude GAIAX from the market. Moreover, the strong aversion by the Japanese consumer to a high density alcohol fuel of any type may prevent commercial success of any new fuel.

Methanol

Methanol fuel is an alternative biofuel for internal combustion and other engines, either in combination with gasoline or independently. Methanol is less expensive to produce sustainably than ethanol fuel, although it is generally more toxic and has lower energy density. For optimizing engine performance and fuel availability, however, a blend of ethanol, methanol and petroleum is likely to be preferable to using any of these alone. Methanol may be made from hydrocarbon or renewable resources, in particular natural gas and biomass respectively. It can also be synthesized from CO_2 (carbon dioxide) and hydrogen. Methanol fuel is currently used by racing cars in many countries but has not seen widespread use otherwise.

Production

Historically, methanol was first produced by destructive distillation (pyrolysis) of wood, resulting in its common English name of wood alcohol.

At present, methanol is usually produced using methane (the chief constituent of natural gas) as a raw material. In China, methanol is made for fuel from coal.

"Biomethanol" may be produced by gasification of organic materials to synthesis gas followed by conventional methanol synthesis. This route can offer methanol production from biomass at efficiencies up to 75%. Widespread production by this route has a proposed potential to offer methanol fuel at a low cost and with benefits to the environment. These production methods, however, are not suitable for small-scale production.

Recently, methanol fuel has been produced using renewable energy and carbon dioxide as a feedstock. Carbon Recycling International, an Icelandic-American company, completed the first commercial scale renewable methanol plant in 2011.

Major Fuel Use

During the OPEC 1973 oil crisis, Reed and Lerner proposed methanol from coal as a proven fuel with well-established manufacturing technology and sufficient resources to replace gasoline. Hagen reviewed prospects for synthesizing methanol from fossil and renewable resources, its use as a fuel, economics, and hazards. Then in 1986, the Swedish Motor Fuel Technology Co. (SBAD) extensively reviewed the use of alcohols and alcohol blends as motor fuels. It reviewed the potential for methanol production from natural gas, very heavy oils, bituminous shales, coals, peat and biomass. In 2005, 2006 Nobel prize winner George A. Olah, G. K. Surya Prakash and Alain Goeppert advocated an entire methanol economy based on energy storage in synthetically produced methanol. The Methanol Institute, the methanol trade industry organization, posts reports and presentations on methanol. Director Gregory Dolan presented the 2008 global methanol fuel industry in China.

On January 26, 2011, the European Union's Directorate-General for Competition approved the Swedish Energy Agency's award of 500 million Swedish kronor (approx. €56M as at January 2011) toward the construction of a 3 billion Swedish kronor (approx. €335M) industrial scale experimental development biofuels plant for production of Biomethanol and BioDME at the Domsjö Fabriker biorefinery complex in Örnsköldsvik, Sweden, using Chemrec's black liquor gasification technology.

Uses

Internal Combustion Engine Fuel

Both methanol and ethanol burn at lower temperatures than gasoline, and both are less volatile, making engine starting in cold weather more difficult. Using methanol as a fuel in spark-ignition engines can offer an increased thermal efficiency and increased power output (as compared to gasoline) due to its high octane rating and high heat of vaporization. However, its low energy content of 19.7 MJ/kg and stoichiometric air-to-fuel ratio of 6.42:1 mean that fuel consumption (on volume or mass bases) will be higher than hydrocarbon fuels. The extra water produced also makes the charge rather wet (similar to hydrogen/oxygen combustion engines) and with the formation of acidic products during combustion, the wearing of valves, valve seats and cylinder might be higher than with hydrocarbon burning. Certain additives may be added to the fuel in order to neutralize these acids.

Methanol, like ethanol, contains soluble and insoluble contaminants. These soluble contaminants, halide ions such as chloride ions, have a large effect on the corrosivity of alcohol fuels. Halide ions increase corrosion in two ways; they chemically attack passivating oxide films on several metals causing pitting corrosion, and they increase the conductivity of the fuel. Increased electrical conductivity promotes electric, galvanic, and ordinary corrosion in the fuel system. Soluble contaminants, such as aluminum hydroxide, itself a product of corrosion by halide ions, clog the fuel system over time.

Methanol is (In automotive terms) hygroscopic, meaning it will absorb water vapor directly from the atmosphere. Because absorbed water dilutes the fuel value of the methanol (although it suppresses engine knock), and may cause phase separation of methanol-gasoline blends, containers of methanol fuels must be kept tightly sealed.

Compared to gasoline, methanol is more tolerant to exhaust gas recirculation (EGR), which improves fuel efficiency of the internal combustion engines utilizing Otto cycle and spark ignition.

An acid, albeit weak, methanol attacks the oxide coating that normally protects the aluminium from corrosion:

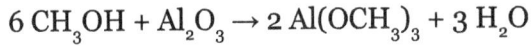

$$6 \, CH_3OH + Al_2O_3 \rightarrow 2 \, Al(OCH_3)_3 + 3 \, H_2O$$

The resulting methoxide salts are soluble in methanol, resulting in a clean aluminium surface, which is readily oxidized by dissolved oxygen. Also, the methanol can act as an oxidizer:

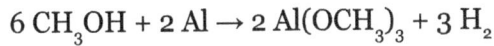

$$6 \, CH_3OH + 2 \, Al \rightarrow 2 \, Al(OCH_3)_3 + 3 \, H_2$$

This reciprocal process effectively fuels corrosion until either the metal is eaten away or the concentration of CH_3OH is negligible. Methanol's corrosivity has been addressed with methanol-compatible materials and fuel additives that serve as corrosion inhibitors.

Organic methanol, produced from wood or other organic materials (bioalcohol), has been suggested as a renewable alternative to petroleum-based hydrocarbons. Low levels of methanol can be used in existing vehicles with the addition of cosolvents and corrosion inhibitors.

Racing

Pure methanol is required by rule to be used in Champcars, Monster Trucks, USAC sprint cars (as well as midgets, modifieds, *etc.*), and other dirt track series, such as World of Outlaws, and Motorcycle Speedway, mainly because, in the event of an accident, methanol does not produce an opaque cloud of smoke. Since the late 1940s, Methanol is also used as the primary fuel ingredient in the powerplants for radio control, control line and free flight model aircraft, cars and trucks; such engines use a platinum filament glow plug that ignites the methanol vapor through a catalytic reaction. Drag racers, mud racers, and heavily modified tractor pullers also use methanol as the primary fuel source. Methanol is required with a supercharged engine in a Top Alcohol Dragster and, until the end of the 2006 season, all vehicles in the Indianapolis 500 had to run on methanol. As a fuel for mud racers, methanol mixed with gasoline and nitrous oxide produces more power than gasoline and nitrous oxide alone.

Beginning in 1965, pure methanol was used widespread in USAC Indy car competition, which at the time included the Indianapolis 500.

Safety was the predominant influence for the adoption of methanol fuel in the United States open-wheel racing categories. Unlike petroleum fires, methanol fires can be extinguished with plain water. A methanol-based fire burns invisibly, unlike gasoline, which burns with a visible flame. If a fire occurs on the track, there is no flame or smoke to obstruct the view of fast approaching drivers, but this can also delay visual detection of the fire and the initiation of fire suppression. A seven-car crash on the second lap of the 1964 Indianapolis 500 resulted in USAC's decision to encourage, and later mandate, the use of methanol. Eddie Sachs and Dave MacDonald died in the crash when their gasoline-fueled cars exploded. The gasoline-triggered fire created a dangerous cloud of thick black smoke that completely blocked the view of the track for oncoming cars. Johnny Rutherford, one of the other drivers involved, drove a methanol-fueled car, which also leaked following the crash. While this car burned from the impact of the first fireball, it formed a much smaller inferno than the gasoline cars, and one that burned invisibly. That testimony, and pressure from *The Indianapolis Star* writer George Moore, led to the switch to alcohol fuel in 1965.

Methanol was used by the CART circuit during its entire campaign. It is also used by many-short track organizations, especially midget, sprint cars and speedway bikes. Pure methanol was used by the IRL from 1996-2006.

In 2006, in partnership with the ethanol industry, the IRL used a mixture of 10% ethanol and 90% methanol as its fuel. Starting in 2007, the IRL switched to "pure" ethanol, E100.

Methanol fuel is also used extensively in drag racing, primarily in the Top Alcohol category, while between 10% and 20% methanol may be used in Top Fuel classes in addition to Nitromethane.

Formula One racing continues to use gasoline as its fuel, but in prewar grand prix racing methanol was often used in the fuel.

Methanol is also used in Monster Truck racing.

Fuel for Model Engines

The earliest model engines for free-flight model aircraft flown before the end of World War II used a 3:1 mix of white gas and heavy viscosity motor oil for the two-stroke spark ignition engines used for the hobby at that time. By 1948, the new glow plug-ignition model engines began to take over the market, requiring the use of methanol fuel to react in a catalytic reaction with the coiled platinum filament in a glow plug for the engine to run, usually using a castor oil-based lubricant contained in the fuel mix at about a 4:1 ratio. The glow-ignition variety of model engine, because it no longer required an onboard battery, ignition coil, ignition points and capacitor that a spark ignition model engine required, saved valuable weight and allowed model aircraft to have better flight performance. In their traditionally popular two-stroke and increasingly popular four-stroke forms, currently produced single cylinder methanol-fueled glow engines are the usual choice for radio controlled aircraft for recreational use, for engine sizes that can range from 0.8 cm^3 (0.049 cu.in.) to as large as 25 to 32 cm^3 (1.5-2.0 cu.in) displacement, and significantly larger displacements for twin and multi-cylinder opposed-cylinder and radial configuration model aircraft engines, many of which are of four-stroke configuration. Most methanol-fueled model engines, especially those made outside North America, can easily be run on so-called *FAI*-specification methanol fuel. Such fuel mixtures can be required by the FAI for certain events in so-called FAI "Class F" international competition, that forbid the use of nitromethane as a glow engine fuel component. In contrast, firms in North America that make methanol-fueled model engines, or who are based outside that continent and have a major market in North America for such miniature powerplants, tend to produce engines that can and often do run best with a certain percentage of nitromethane in the fuel, which when used can be as little as 5% to 10% of volume, and can be as much as 25 to 30% of the total fuel volume.

Cooking

Methanol is used as a cooking fuel in China and its use in India is growing. Its stove and canister need no regulators or pipes.

Toxicity

Methanol occurs naturally in the human body and in some fruits, but is poisonous in high concentration. Ingestion of 10 ml can cause blindness and 60-100 ml can be fatal if the condition is untreated. Like many volatile chemicals, methanol does not have to be swallowed to be dangerous since the liquid can be absorbed through the skin, and the vapors through the lungs. Methanol fuel is much safer when blended with ethanol, even at relatively low ethanol percentages.

US maximum allowed exposure in air (40 h/week) is 1900 mg/m³ for ethanol, 900 mg/m³ for gasoline, and 1260 mg/m³ for methanol. However, it is much less volatile than gasoline, and therefore has lower evaporative emissions, producing a lower exposure risk for an equivalent spill. While methanol offers somewhat different toxicity exposure pathways, the effective toxicity is no worse than those of benzene or gasoline, and methanol poisoning is far easier to treat successfully. One substantial concern is that methanol poisoning generally must be treated while it is still asymptomatic for full recovery.

Inhalation risk is mitigated by a characteristic pungent odor. At concentrations greater than 2,000 ppm (0.2%) it is generally quite noticeable, however lower concentrations may remain undetected while still being potentially toxic over longer exposures, and may still present a fire/explosion hazard. Again, this is similar to gasoline and ethanol; standard safety protocols exist for methanol and are very similar to those for gasoline and ethanol.

Use of methanol fuel reduces the exhaust emissions of certain hydrocarbon-related toxins such as benzene and 1,3 butadiene, and dramatically reduces long term groundwater pollution caused by fuel spills. Unlike benzene-family fuels, methanol will rapidly and non-toxically biodegrade with no long-term harm to the environment as long as it is sufficiently diluted.

Fire Safety

Methanol is far more difficult to ignite than gasoline and burns about 60% slower. A methanol fire releases energy at around 20% of the rate of a gasoline fire, resulting in a much cooler flame. This results in a much less dangerous fire that is easier to contain with proper protocols. Unlike gasoline fires, water is acceptable and even preferred as a fire suppressant for methanol fires, since this both cools the fire and rapidly dilutes the fuel below the concentration where it will maintain self-flammability. These facts mean that, as a vehicle fuel, methanol has great safety advantages over gasoline. Ethanol shares many of these same advantages.

Since methanol vapor is heavier than air, it will linger close to the ground or in a pit unless there is good ventilation, and if the concentration of methanol is above 6.7% in air it can be lit by a spark and will explode above 54 °F / 62 °C. Once ablaze, an undiluted methanol fire gives off very little visible light, making it potentially very hard to see the fire or even estimate its size in bright daylight, although in the vast majority of cases, existing pollutants or flammables in the fire (such as tires or asphalt) will color and enhance the visibility of the fire. Ethanol, natural gas, hydrogen, and other existing fuels offer similar fire-safety challenges, and standard safety and firefighting protocols exist for all such fuels.

Post-accident environmental damage mitigation is facilitated by the fact that low-concentration methanol is biodegradable, of low toxicity, and non-persistent in the environment. Post-fire cleanup often merely requires large additional amounts of water to dilute the spilled methanol followed by vacuuming or absorption recovery of the fluid. Any methanol that unavoidably escapes into the environment will have little long-term impact, and with sufficient dilution will rapidly biodegrade with little to no environmental damage due to toxicity. A methanol spill that combines with an existing gasoline spill can cause the mixed methanol/gasoline spill to persist about 30% to 35% longer than the gasoline alone would have done.

Use

United States

The State of California ran an experimental program from 1980 to 1990 that allowed anyone to convert a gasoline vehicle to 85% methanol with 15% additives of choice. Over 500 vehicles were converted to high compression and dedicated use of the 85/15 methanol and ethanol.

In 1982 the big three were each given $5,000,000 for design and contracts for 5,000 vehicles to be bought by the State. It was an early use of low-compression flexible-fuel vehicles.

In 2005, California's Governor, Arnold Schwarzenegger, stopped the use of methanol to join the expanding use of ethanol driven by producers of corn. In 2007 ethanol was priced at 3 to 4 dollars per gallon (0.8 to 1.05 dollars per liter) at the pump, while methanol made from natural gas remains at 47 cents per gallon (12.5 cents per liter) in bulk, not at the pump.

Presently there are no operating gas stations in California supplying methanol in their pumps. Rep. Eliot Engel [D-NY17] has introduced "An Open Fuel Standard" Act in Congress: "To require automobile manufacturers to ensure that not less than 80 percent of the automobiles manufactured or sold in the United States by each such manufacturer to operate on fuel mixtures containing 85 percent ethanol, 85 percent methanol, or biodiesel".

European Union

The amended Fuel Quality Directive adopted in 2009 allows up to 3% v/v blend-in of methanol in petrol.

Brazil

A drive to add an appreciable percentage of methanol to gasoline got very close to implementation in Brazil, following a pilot test set up by a group of scientists involving blending gasoline with methanol between 1989 and 1992. The larger-scale pilot experiment that was to be conducted in São Paulo was vetoed at the last minute by the city's mayor, out of concern for the health of gas station workers, who would not be expected to follow safety precautions. As of 2006, the idea has not resurfaced.

India

Niti Aayog, The planning commission of India on 3rd August 2018 announced that if feasible, passenger vehicles will run on 15% Methanol mixed fuel. At present, vehicles in India use up to 10% ethanol-blended fuel. If approved by the government it will cut monthly fuel costs by 10%. In India ethanol costs Rs 42 a litre, while the price of methanol has been estimated at less than Rs 20 a litre.

Ethanol

Ethanol, which is sometimes known as ethyl alcohol, is a kind of alcohol derived from corn, sugarcane, and grain or indirectly from paper waste. It's also the main type of alcohol in most alcoholic beverages obtained as a result of fermentation of a mash of grains (gin, vodka, and whiskey) or sugarcane (rums). It's also a source of fuel commonly blended with gasoline to oxygenate the fuel at the gas pump. Ethanol fuel can also be used on its own to power vehicles.

Ethanol is more common in our lives than you may think. After all, any alcoholic beverage you can drink comprises of Ethanol. It is known by many different names such as Ethyl alcohol, pure alcohol and grain alcohol. It is regarded as an alternative form of fuel that has gained much popularity for a number of reasons.

The most common use of Ethanol fuel is by blending it with gasoline. Doing so creates a mix that releases fewer emissions into the environment and is considered cleaner in nature. It also keeps the car in a better shape by increasing the octane rating of the fuel. All in all, it is accepted by the people, governments and car companies for the many benefits it provides.

In case you were wondering, Ethanol does not occur naturally in any eco-system. It is produced through the processes of fermentation and distillation. While the energy based use of Ethanol fuel is new, it has been part of our lives for a very long time. Fermenting sugar creates Ethanol – knowledge used by our forefathers. These days, it comes from crops and plants that are rich in sugar or have the ability to be converted into cellulose and starch. Sugarcane, barley, sugar beets, wheat and corn are commonly used for production.

Chemistry

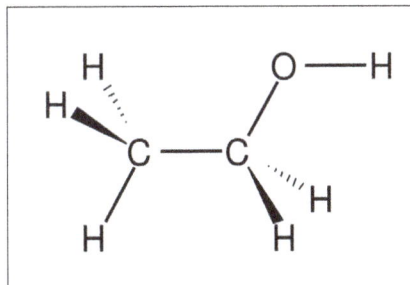

Structure of ethanol molecule. All bonds are single bonds.

During ethanol fermentation, glucose and other sugars in the corn (or sugarcane or other crops) are converted into ethanol and carbon dioxide.

$$C_6H_{12}O_6 \rightarrow 2\ C_2H_5OH + 2\ CO_2 + heat$$

Ethanol fermentation is not 100% selective with side products such as acetic acid and glycols. They are mostly removed during ethanol purification. Fermentation takes place in an aqueous solution. The resulting solution has an ethanol content of around 15%. Ethanol is subsequently isolated and purified by a combination of adsorption and distillation.

During combustion, ethanol reacts with oxygen to produce carbon dioxide, water, and heat:

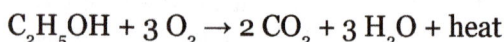

$$C_2H_5OH + 3\ O_2 \rightarrow 2\ CO_2 + 3\ H_2O + heat$$

Starch and cellulose molecules are strings of glucose molecules. It is also possible to generate ethanol out of cellulosic materials. That, however, requires a pretreatment that splits the cellulose into glucose molecules and other sugars that subsequently can be fermented. The resulting product is called cellulosic ethanol, indicating its source.

Ethanol is also produced industrially from ethylene by hydration of the double bond in the presence of a catalyst and high temperature.

$$C_2H_4 + H_2O \rightarrow C_2H_5OH$$

Most ethanol is produced by fermentation.

Sources

Sugar cane harvest.

Cornfield.

About 5% of the ethanol produced in the world in 2003 was actually a petroleum product. It is made by the catalytic hydration of ethylene with sulfuric acid as the catalyst. It can also be obtained via ethylene or acetylene, from calcium carbide, coal, oil gas, and other sources. Two million short tons (1,786,000 long tons; 1,814,000 t) of petroleum-derived ethanol are produced annually. The principal suppliers are plants in the United States, Europe, and South Africa. Petroleum derived ethanol (synthetic ethanol) is chemically identical to bio-ethanol and can be differentiated only by radiocarbon dating.

Bio-ethanol is usually obtained from the conversion of carbon-based feedstock. Agricultural feedstocks are considered renewable because they get energy from the sun using photosynthesis, provided that all minerals required for growth (such as nitrogen and phosphorus) are returned to the land. Ethanol can be produced from a variety of feedstocks such as sugar cane, bagasse, miscanthus, sugar beet, sorghum, grain, switchgrass, barley, hemp, kenaf, potatoes, sweet potatoes, cassava, sunflower, fruit, molasses, corn, stover, grain, wheat, straw, cotton, other biomass, as well as many types of cellulose waste and harvesting, whichever has the best well-to-wheel assessment.

An alternative process to produce bio-ethanol from algae is being developed by the company Algenol. Rather than grow algae and then harvest and ferment it, the algae grow in sunlight and produce ethanol directly, which is removed without killing the algae. It is claimed the process can produce 6,000 U.S. gallons per acre (5,000 imperial gallons per acre; 56,000 liters per hectare) per year compared with 400 US gallons per acre (330 imp gal/acre; 3,700 L/ha) for corn production.

Switchgrass.

Currently, the first generation processes for the production of ethanol from corn use only a small part of the corn plant: the corn kernels are taken from the corn plant and only the starch, which represents about 50% of the dry kernel mass, is transformed into ethanol. Two types of second generation processes are under development. The first type uses enzymes and yeast fermentation to convert the plant cellulose into ethanol while the second type uses pyrolysis to convert the whole plant to either a liquid bio-oil or a syngas. Second generation processes can also be used with plants such as grasses, wood or agricultural waste material such as straw.

Production

Although there are various ways ethanol fuel can be produced, the most common way is via fermentation.

The basic steps for large-scale production of ethanol are: microbial (yeast) fermentation of sugars, distillation, dehydration, and denaturing (optional). Prior to fermentation, some crops require saccharification or hydrolysis of carbohydrates such as cellulose and starch into sugars. Saccharification of cellulose is called cellulolysis. Enzymes are used to convert starch into sugar.

Fermentation

Ethanol is produced by microbial fermentation of the sugar. Microbial fermentation currently only works directly with sugars. Two major components of plants, starch and cellulose, are both made of sugars—and can, in principle, be converted to sugars for fermentation. Currently, only the sugar (e.g., sugar cane) and starch (e.g., corn) portions can be economically converted. There is much activity in the area of cellulosic ethanol, where the cellulose part of a plant is broken down to sugars and subsequently converted to ethanol.

Distillation

Ethanol plant.

For the ethanol to be usable as a fuel, the yeast solids and the majority of the water must be removed. After fermentation, the mash is heated so that the ethanol evaporates. This process, known as distillation, separates the ethanol, but its purity is limited to 95–96% due to the formation of a low-boiling water-ethanol azeotrope with maximum (95.6% m/m (96.5% v/v) ethanol and 4.4% m/m (3.5% v/v) water). This mixture is called hydrous ethanol and can be used as a fuel alone, but unlike anhydrous ethanol, hydrous ethanol is not miscible in all ratios with gasoline, so the water fraction is typically removed in further treatment to burn in combination with gasoline in gasoline engines.

Ethanol plant.

Dehydration

There are three dehydration processes to remove the water from an azeotropic ethanol/water mixture. The first process, used in many early fuel ethanol plants, is called azeotropic distillation and consists of adding benzene or cyclohexane to the mixture. When these components are added to the mixture, it forms a heterogeneous azeotropic mixture in vapor–liquid-liquid equilibrium, which when distilled produces anhydrous ethanol in the column bottom, and a vapor mixture of water, ethanol, and cyclohexane/benzene.

When condensed, this becomes a two-phase liquid mixture. The heavier phase, poor in the entrainer (benzene or cyclohexane), is stripped of the entrainer and recycled to the feed—while the lighter phase, with condensate from the stripping, is recycled to the second column. Another early method, called extractive distillation, consists of adding a ternary component that increases ethanol's relative volatility. When the ternary mixture is distilled, it produces anhydrous ethanol on the top stream of the column.

With increasing attention being paid to saving energy, many methods have been proposed that avoid distillation altogether for dehydration. Of these methods, a third method has emerged and has been adopted by the majority of modern ethanol plants. This new process uses molecular sieves to remove water from fuel ethanol. In this process, ethanol vapor under pressure passes through a bed of molecular sieve beads. The bead's pores are sized to allow adsorption of water while excluding ethanol. After a period of time, the bed is regenerated under vacuum or in the flow of inert atmosphere (e.g. N_2) to remove the adsorbed water. Two beds are often used so that one is available to adsorb water while the other is being regenerated. This dehydration technology can account for energy saving of 3,000 btus/gallon (840 kJ/L) compared to earlier azeotropic distillation.

Recent research has demonstrated that complete dehydration prior to blending with gasoline is not always necessary. Instead, the azeotropic mixture can be blended directly with gasoline so that liquid-liquid phase equilibrium can assist in the elimination of water. A two-stage counter-current setup of mixer-settler tanks can achieve complete recovery of ethanol into the fuel phase, with minimal energy consumption.

Post-production Water Issues

Ethanol is hygroscopic, meaning it absorbs water vapor directly from the atmosphere. Because absorbed water dilutes the fuel value of the ethanol and may cause phase separation of ethanol-gasoline blends (which causes engine stall), containers of ethanol fuels must be kept tightly sealed. This high miscibility with water means that ethanol cannot be efficiently shipped through modern pipelines, like liquid hydrocarbons, over long distances.

The fraction of water that an ethanol-gasoline fuel can contain without phase separation increases with the percentage of ethanol. For example, E30 can have up to about 2% water. If there is more than about 71% ethanol, the remainder can be any proportion of water or gasoline and phase separation does not occur. The fuel mileage declines with increased water content. The increased solubility of water with higher ethanol content permits E30 and hydrated ethanol to be put in the same tank since any combination of them always results in a single phase. Somewhat less water is tolerated at lower temperatures. For E10 it is about 0.5% v/v at 21° C and decreases to about 0.23% v/v at −34 °C .

Consumer Production Systems

While biodiesel production systems have been marketed to home and business users for many years, commercialized ethanol production systems designed for end-consumer use have lagged in the marketplace. In 2008, two different companies announced home-scale ethanol production systems. The AFS125 Advanced Fuel System from Allard Research and Development is capable of producing both ethanol and biodiesel in one machine, while the E-100 MicroFueler from E-Fuel Corporation is dedicated to ethanol only.

Engines

Fuel Economy

Ethanol contains approx. 34% less energy per unit volume than gasoline, and therefore in theory, burning pure ethanol in a vehicle reduces miles per US gallon 34%, given the same fuel economy, compared to burning pure gasoline. However, since ethanol has a higher octane rating, the engine can be made more efficient by raising its compression ratio. Using a variable geometry or twin scroll turbocharger, the compression ratio can be optimized for the fuel, making fuel economy almost constant for any blend.

For E10 (10% ethanol and 90% gasoline), the effect is small (~3%) when compared to conventional gasoline, and even smaller (1–2%) when compared to oxygenated and reformulated blends. For E85 (85% ethanol), the effect becomes significant. E85 produces lower mileage than gasoline, and requires more frequent refueling. Actual performance may vary depending on the vehicle. Based on EPA tests for all 2006 E85 models, the average fuel economy for E85 vehicles was 25.56% lower than unleaded gasoline. The EPA-rated mileage of current United States flex-fuel vehicles should be considered when making price comparisons, but E85 is a high performance fuel, with an octane rating of about 94–96, and should be compared to premium.

Cold Start during the Winter

The Brazilian 2008 Honda Civic flex-fuel has outside direct access to the secondary reservoir gasoline tank in the front right side, the corresponding fuel filler door is shown by the arrow.

High ethanol blends present a problem to achieve enough vapor pressure for the fuel to evaporate and spark the ignition during cold weather (since ethanol tends to increase fuel enthalpy of vaporization). When vapor pressure is below 45 kPa starting a cold engine becomes difficult. To avoid this problem at temperatures below 11 °C (52 °F), and to reduce ethanol higher emissions during cold weather, both the US and the European markets adopted E85 as the maximum blend to be used in their flexible fuel vehicles, and they are optimized to run at such a blend. At places with harsh cold weather, the ethanol blend in the US has a seasonal reduction to E70 for these very cold regions, though it is still sold as E85. At places where temperatures fall below –12 °C (10 °F) during the winter, it is recommended to install an engine heater system, both for gasoline and E85 vehicles. Sweden has a similar seasonal reduction, but the ethanol content in the blend is reduced to E75 during the winter months.

Brazilian flex fuel vehicles can operate with ethanol mixtures up to E100, which is hydrous ethanol (with up to 4% water), which causes vapor pressure to drop faster as compared to E85 vehicles. As a result, Brazilian flex vehicles are built with a small secondary gasoline reservoir located near the engine. During a cold start pure gasoline is injected to avoid starting problems at low temperatures. This provision is particularly necessary for users of Brazil's southern and central regions, where temperatures normally drop below 15 °C (59 °F) during the winter. An improved flex engine generation was launched in 2009 that eliminates the need for the secondary gas storage tank. In March 2009 Volkswagen do Brasil launched the Polo E-Flex, the first Brazilian flex fuel model without an auxiliary tank for cold start.

Fuel Mixtures

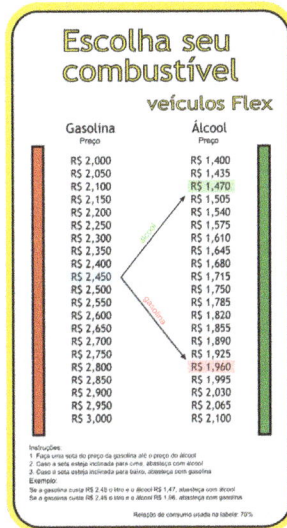

Hydrated ethanol × gasoline type C price table for use in Brazil

In many countries cars are mandated to run on mixtures of ethanol. All Brazilian light-duty vehicles are built to operate for an ethanol blend of up to 25% (E25), and since 1993 a federal law requires mixtures between 22% and 25% ethanol, with 25% required as of mid July 2011. In the United States all light-duty vehicles are built to operate normally with an ethanol blend of 10% (E10). At the end of 2010 over 90 percent of all gasoline sold in the U.S. was blended with ethanol. In January 2011 the U.S. Environmental Protection Agency (EPA) issued a waiver to authorize up to 15% of ethanol blended with gasoline (E15) to be sold only for cars and light pickup trucks with a model year of 2001 or newer.

Beginning with the model year 1999, an increasing number of vehicles in the world are manufactured with engines that can run on any fuel from 0% ethanol up to 100% ethanol without modification. Many cars and light trucks (a class containing minivans, SUVs and pickup trucks) are designed to be flexible-fuel vehicles using ethanol blends up to 85% (E85) in North America and Europe, and up to 100% (E100) in Brazil. In older model years, their engine systems contained alcohol sensors in the fuel and/or oxygen sensors in the exhaust that provide input to the engine control computer to adjust the fuel injection to achieve stochiometric (no residual fuel or free oxygen in the exhaust) air-to-fuel ratio for any fuel mix. In newer models, the alcohol sensors have been removed, with the computer using only oxygen and airflow sensor feedback to estimate alcohol content. The engine control computer can also adjust (advance) the ignition timing to achieve a higher output without pre-ignition when it predicts that higher alcohol percentages are present in the fuel being burned. This method is backed up by advanced knock sensors – used in most high performance gasoline engines regardless of whether they are designed to use ethanol or not – that detect pre-ignition and detonation.

Other Engine Configuration

ED95 Engines

Since 1989 there have also been ethanol engines based on the diesel principle operating in Sweden. They are used primarily in city buses, but also in distribution trucks and waste collectors. The engines, made by Scania, have a modified compression ratio, and the fuel (known as ED95) used is a mix of 93.6% ethanol and 3.6% ignition improver, and 2.8% denaturants. The ignition improver makes it possible for the fuel to ignite in the diesel combustion cycle. It is then also possible to use the energy efficiency of the diesel principle with ethanol. These engines have been used in the United Kingdom by Reading Buses but the use of bioethanol fuel is now being phased out.

Dual-fuel Direct-injection

A 2004 MIT study and an earlier paper published by the Society of Automotive Engineers identified a method to exploit the characteristics of fuel ethanol substantially more efficiently than mixing it with gasoline. The method presents the possibility of leveraging the use of alcohol to achieve definite improvement over the cost-effectiveness of hybrid electric. The improvement consists of using dual-fuel direct-injection of pure alcohol (or the azeotrope or E85) and gasoline, in any ratio up to 100% of either, in a turbocharged, high compression-ratio, small-displacement engine having performance similar to an engine having twice the displacement. Each fuel is carried separately, with a much smaller tank for alcohol. The high-compression (for higher efficiency) engine runs on ordinary gasoline under low-power cruise conditions. Alcohol is directly injected into the cylinders (and the gasoline injection simultaneously reduced) only when necessary to suppress 'knock' such as when significantly accelerating. Direct cylinder injection raises the already high octane rating of ethanol up to an effective 130. The calculated over-all reduction of gasoline use and CO_2 emission is 30%. The consumer cost payback time shows a 4:1 improvement over turbo-diesel and a 5:1 improvement over hybrid. The problems of water absorption into pre-mixed gasoline (causing phase separation), supply issues of multiple mix ratios and cold-weather starting are also avoided.

Increased Thermal Efficienc

In a 2008 study, complex engine controls and increased exhaust gas recirculation allowed a compression ratio of 19.5 with fuels ranging from neat ethanol to E50. Thermal efficiency up to approximately that for a diesel was achieved. This would result in the fuel economy of a neat ethanol vehicle to be about the same as one burning gasoline.

Fuel Cells Powered by an Ethanol Reformer

In June 2016, Nissan announced plans to develop fuel cell vehicles powered by ethanol rather than hydrogen, the fuel of choice by the other car manufacturers that have developed and commercialized fuel cell vehicles, such as the Hyundai Tucson FCEV, Toyota Mirai, and Honda FCX Clarity. The main advantage of this technical approach is that it would be cheaper and easier to deploy the fueling infrastructure than setting up the one required to deliver hydrogen at high pressures, as each hydrogen fueling station cost US$1 million to US$2 million to build.

Nissan plans to create a technology that uses liquid ethanol fuel as a source to generate hydrogen within the vehicle itself. The technology uses heat to reform ethanol into hydrogen to feed what is known as a solid oxide fuel cell (SOFC). The fuel cell generates electricity to supply power to the electric motor driving the wheels, through a battery that handles peak power demands and stores regenerated energy. The vehicle would include a tank for a blend of water and ethanol, which is fed into an onboard reformer that splits it into pure hydrogen and carbon dioxide. According to Nissan, the liquid fuel could be an ethanol-water blend at a 55:45 ratio. Nissan expects to commercialize its technology by 2020.

Environment

Energy Balance

Energy balance		
Country	Type	Energy balance
United States	Corn ethanol	1.3
Germany	Biodiesel	2.5
Brazil	Sugarcane ethanol	8
United States	Cellulosic ethanol	2–36

All biomass goes through at least some of these steps: It needs to be grown, collected, dried, fermented, distilled, and burned. All of these steps require resources and an infrastructure. The total amount of energy input into the process compared to the energy released by burning the resulting ethanol fuel is known as the energy balance (or "energy returned on energy invested"). Figures compiled in a 2007 report by *National Geographic Magazine* point to modest results for corn ethanol produced in the US: one unit of fossil-fuel energy is required to create 1.3 energy units from the resulting ethanol. The energy balance for sugarcane ethanol produced in Brazil is more favorable, with one unit of fossil-fuel energy required to create 8 from the ethanol. Energy balance estimates are not easily produced, thus numerous such reports have been generated that are contradictory. For instance, a separate survey reports that production of ethanol from sugarcane, which requires a tropical climate to grow productively, returns from 8 to 9 units of energy for each unit expended, as compared to corn, which only returns about 1.34 units of fuel energy for each unit of energy expended. A 2006 University of California Berkeley study, after analyzing six separate studies, concluded that producing ethanol from corn uses much less petroleum than producing gasoline.

Carbon dioxide, a greenhouse gas, is emitted during fermentation and combustion. This is canceled out by the greater uptake of carbon dioxide by the plants as they grow to produce the biomass. When compared to gasoline, depending on the production method, ethanol releases less greenhouse gases.

Air Pollution

Compared with conventional unleaded gasoline, ethanol is a particulate-free burning fuel source that combusts with oxygen to form carbon dioxide, carbon monoxide, water and aldehydes. The Clean Air Act requires the addition of oxygenates to reduce carbon monoxide emissions in the United States. The additive MTBE is currently being phased out due to ground water contamination, hence ethanol becomes an attractive alternative additive. Current production methods include air pollution from the manufacturer of macronutrient fertilizers such as ammonia.

A study by atmospheric scientists at Stanford University found that E85 fuel would increase the risk of air pollution deaths relative to gasoline by 9% in Los Angeles, US: a very large, urban, car-based metropolis that is a worst-case scenario. Ozone levels are significantly increased, thereby increasing photochemical smog and aggravating medical problems such as asthma.

Brazil burns significant amounts of ethanol biofuel. Gas chromatograph studies were performed of ambient air in São Paulo, Brazil, and compared to Osaka, Japan, which does not burn ethanol fuel. Atmospheric Formaldehyde was 160% higher in Brazil, and Acetaldehyde was 260% higher.

Carbon Dioxide

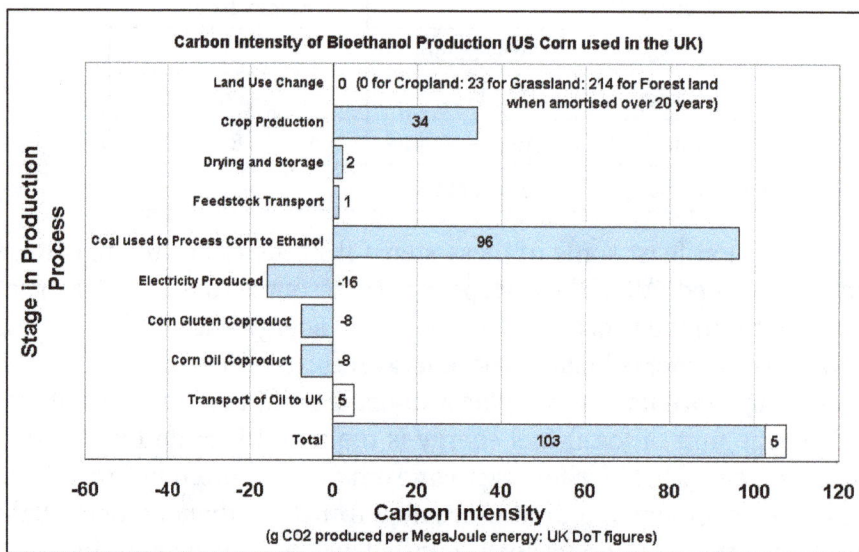

Carbon Intensity of Bioethanol Production (US Corn used in the UK)

Stage in Production Process	Carbon intensity (g CO2 produced per MegaJoule energy: UK DoT figures)
Land Use Change	0 (0 for Cropland: 23 for Grassland: 214 for Forest land when amortised over 20 years)
Crop Production	34
Drying and Storage	2
Feedstock Transport	1
Coal used to Process Corn to Ethanol	96
Electricity Produced	-16
Corn Gluten Coproduct	-8
Corn Oil Coproduct	-8
Transport of Oil to UK	5
Total	103 5

UK government calculation of carbon intensity of corn bioethanol grown in the US and burnt in the UK.

The calculation of exactly how much carbon dioxide is produced in the manufacture of bioethanol is a complex and inexact process, and is highly dependent on the method by which the ethanol is produced and the assumptions made in the calculation. A calculation should include:

- The cost of growing the feedstock.
- The cost of transporting the feedstock to the factory.
- The cost of processing the feedstock into bioethanol.

Such a calculation may or may not consider the following effects:

- The cost of the change in land use of the area where the fuel feedstock is grown.

- The cost of transportation of the bioethanol from the factory to its point of use.

- The efficiency of the bioethanol compared with standard gasoline.

- The amount of carbon dioxide produced at the tail pipe.

- The benefits due to the production of useful bi-products, such as cattle feed or electricity.

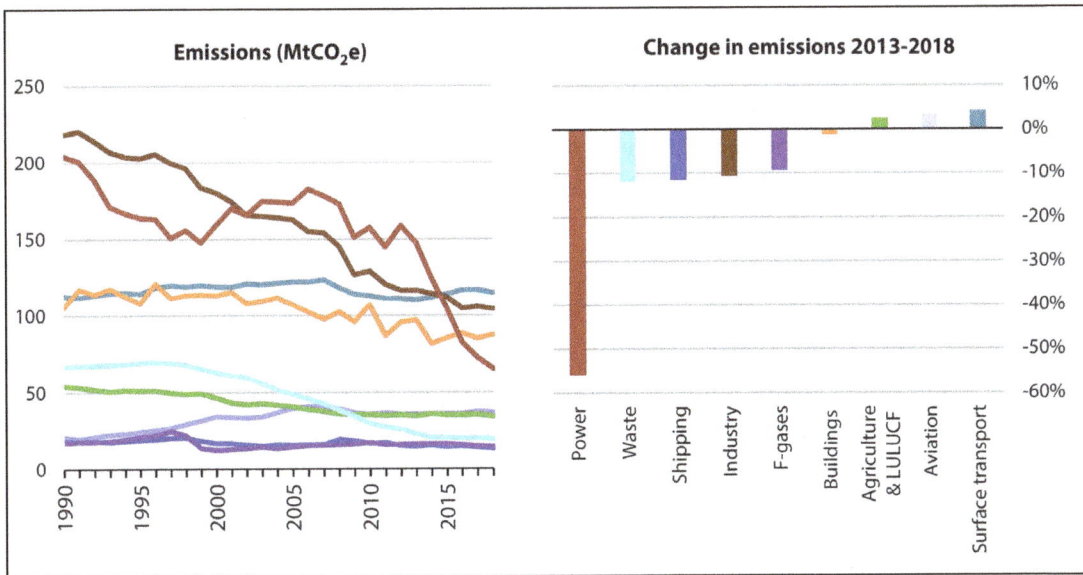

Graph of UK figures for the carbon intensity of bioethanol and fossil fuels. This graph assumes that all bioethanols are burnt in their country of origin and that previously existing cropland is used to grow the feedstock.

The graph shows figures calculated by the UK government for the purposes of the Renewable transport fuel obligation.

The January 2006 Science article from UC Berkeley's ERG, estimated reduction from corn ethanol in GHG to be 13% after reviewing a large number of studies. In a correction to that article released shortly after publication, they reduce the estimated value to 7.4%. A National Geographic Magazine overview article puts the figures at 22% less CO_2 emissions in production and use for corn ethanol compared to gasoline and a 56% reduction for cane ethanol. Carmaker Ford reports a 70% reduction in CO_2 emissions with bioethanol compared to petrol for one of their flexible-fuel vehicles.

An additional complication is that production requires tilling new soil which produces a one-off release of GHG that it can take decades or centuries of production reductions in GHG emissions to equalize. As an example, converting grass lands to corn production for ethanol takes about a century of annual savings to make up for the GHG released from the initial tilling.

Change in Land Use

Large-scale farming is necessary to produce agricultural alcohol and this requires substantial amounts of cultivated land. University of Minnesota researchers report that if all corn grown in the U.S. were used to make ethanol it would displace 12% of current U.S. gasoline consumption. There are claims that land for ethanol production is acquired through deforestation, while others have observed that areas currently supporting forests are usually not suitable for growing crops. In any case, farming may involve a decline in soil fertility due to reduction of organic matter, a decrease in water availability and quality, an increase in the use of pesticides and fertilizers, and potential dislocation of local communities. New technology enables farmers and processors to increasingly produce the same output using less inputs.

Cellulosic ethanol production is a new approach that may alleviate land use and related concerns. Cellulosic ethanol can be produced from any plant material, potentially doubling yields, in an effort to minimize conflict between food needs vs. fuel needs. Instead of utilizing only the starch by-products from grinding wheat and other crops, cellulosic ethanol production maximizes the use of all plant materials, including gluten. This approach would have a smaller carbon footprint because the amount of energy-intensive fertilisers and fungicides remain the same for higher output of usable material. The technology for producing cellulosic ethanol is currently in the commercialization stage.

Using Biomass for Electricity Instead of Ethanol

Converting biomass to electricity for charging electric vehicles may be a more "climate-friendly" transportation option than using biomass to produce ethanol fuel, according to an analysis published in Science in May 2009 Researchers continue to search for more cost-effective developments in both cellulosic ethanol and advanced vehicle batteries.

Health Costs of Ethanol Emissions

For each billion ethanol-equivalent gallons of fuel produced and combusted in the US, the combined climate-change and health costs are $469 million for gasoline, $472–952 million for corn ethanol depending on biorefinery heat source (natural gas, corn stover, or coal) and technology, but only $123–208 million for cellulosic ethanol depending on feedstock (prairie biomass, Miscanthus, corn stover, or switchgrass).

Efficienc of Common Crops

As ethanol yields improve or different feedstocks are introduced, ethanol production may become more economically feasible in the US. Currently, research on improving ethanol yields from each unit of corn is underway using biotechnology. Also, as long as oil prices remain high, the economical use of other feedstocks, such as cellulose, become viable. By-products such as straw or wood chips can be converted to ethanol. Fast growing species like switchgrass can be grown on land not suitable for other cash crops and yield high levels of ethanol per unit area.

Crop	Annual yield (Liters/ hectare, US gal/acre)	Greenhouse-gas savings vs. petrol	Comments
Sugar cane	6800–8000 L/ha, 727–870 g/acre	87%–96%	Long-season annual grass. Used as feedstock for most bioethanol produced in Brazil. Newer processing plants burn residues not used for ethanol to generate electricity. Grows only in tropical and subtropical climates.
Mis-canthus	7300 L/ha, 780 g/acre	37%–73%	Low-input perennial grass. Ethanol production depends on development of cellulosic technology.

Switch-grass	3100–7600 L/ha, 330–810 g/acre	37%–73%	Low-input perennial grass. Ethanol production depends on development of cellulosic technology. Breeding efforts underway to increase yields. Higher biomass production possible with mixed species of perennial grasses.
Poplar	3700–6000 L/ha, 400–640 g/acre	51%–100%	Fast-growing tree. Ethanol production depends on development of cellulosic technology. Completion of genomic sequencing project will aid breeding efforts to increase yields.
Sweet sorghum	2500–7000 L/ha, 270–750 g/acre	No data	Low-input annual grass. Ethanol production possible using existing technology. Grows in tropical and temperate climates, but highest ethanol yield estimates assume multiple crops per year (possible only in tropical climates). Does not store well.
Corn	3100–4000 L/ha, 330–424 g/acre	10%–20%	High-input annual grass. Used as feedstock for most bioethanol produced in USA. Only kernels can be processed using available technology; development of commercial cellulosic technology would allow stover to be used and increase ethanol yield by 1,100 – 2,000 litres/ha.

Reduced Petroleum Imports and Costs

One rationale given for extensive ethanol production in the U.S. is its benefit to energy security, by shifting the need for some foreign-produced oil to domestically produced energy sources. Production of ethanol requires significant energy, but current U.S. production derives most of that energy from coal, natural gas and other sources, rather than oil. Because 66% of oil consumed in the U.S. is imported, compared to a net surplus of coal and just 16% of natural gas (figures from 2006), the displacement of oil-based fuels to ethanol produces a net shift from foreign to domestic U.S. energy sources.

According to a 2008 analysis by Iowa State University, the growth in US ethanol production has caused retail gasoline prices to be US $0.29 to US $0.40 per gallon lower than would otherwise have been the case.

Motorsport

Leon Duray qualified third for the 1927 Indianapolis 500 auto race with an ethanol-fueled car. The IndyCar Series adopted a 10% ethanol blend for the 2006 season, and a 98% blend in 2007.

In drag racing, there are Top Alcohol classes for dragsters and funny cars since the 1970s. The American Le Mans Series sports car championship introduced E10 in the 2007 season to replace pure gasoline. In the 2008 season, E85 was allowed in the GT class and teams began switching to it.

In 2011, the three national NASCAR stock car series mandated a switch from gasoline to E15, a blend of Sunoco GTX unleaded racing fuel and 15% ethanol.

Australia's V8 Supercar championship uses Shell E85 for its racing fuel.

Stock Car Brasil Championship runs on neat ethanol, E100. Ethanol fuel may also be utilized as a rocket fuel. As of 2010, small quantities of ethanol are used in lightweight rocket-racing aircraft.

Replacement Cooking Fuel

Project Gaia is a U.S. non-governmental, non-profit organization involved in the creation of a commercially viable household market for alcohol-based fuels in Ethiopia and other countries in the developing world. The project considers alcohol fuels to be a solution to fuel shortages, environmental damage, and public health issues caused by traditional cooking in the developing world. Targeting poor and marginalized communities that face health issues from cooking over polluting fires, Gaia currently works in Ethiopia, Nigeria, Brazil, Haiti, and Madagascar, and is in the planning stage of projects in several other countries.

Research

Ethanol plant in Turner County.

Ethanol research focuses on alternative sources, novel catalysts and production processes. INEOS produced ethanol from vegetative material and wood waste. The bacterium E.coli when genetically engineered with cow rumen genes and enzymes can produce ethanol from corn stover. Other potential feedstocks are municipal waste, recycled products, rice hulls, sugarcane bagasse, wood chips, switchgrass and carbon dioxide.

Advantages of Ethanol Fuel

Cost Effective

Ethanol fuel is the least expensive energy source since virtually every country has the capability to produce it. Corn, sugar cane or grain grows in almost every country which makes the production economical compared to fossil fuels. Fossils fuels can play against the economy of most countries, especially, developing countries that have no capacity to explore them. It, thus, makes sense for these growing economies to dwell on the production of ethanol fuel to dial back on the dependence of fossil fuel in order to save revenue.

Ecologically Effective

One striking advantage of ethanol over other fuel sources is that it does not cause pollution to the environment. Using ethanol fuel to power automobiles results in significantly low levels of toxins in the environment. On numerous occasions, ethanol is converted to fuel by blending with gasoline. Specifically, ethanol to gasoline ration of 85:15. The little composition of gasoline acts as an igniter, while ethanol takes up the rest of the tasks. This ratio of ethanol to gasoline minimizes the emission of greenhouse gases to the environment since it burns cleanly compared to pure gasoline.

Minimizes Global Warming

Global warming is caused by relentless emission of dangerous greenhouse gases emanation from use of fossil fuels (oil, natural gas, and coal). The effects of global warming are catastrophic including changes in weather patterns, rising sea levels, and excessive heat. Combustion of ethanol fuel only releases carbon dioxide and water. The carbon dioxide released is ineffective regarding environment degradation.

Easily Accessible

Since ethanol is a biofuel, it is easily accessible to virtually everyone. Biofuel means energy derived from plants like sugarcane, grains, and corn. All tropical climates support growth of sugarcane. Grain and corn grow in every country. In fact, corn is a staple food in most countries in Africa.

Minimizes Dependence on Fossil Fuels

Harnessing of fuel from corn or biomass is an economical way to sustain any economy and prevent it from over-reliance on importation of fossil fuels like oil, and gas. Embracing ethanol fuel can save a country a lot of money that can be plowed back into the economy. Since ethanol is domestically produced, from domestically grown crops, it help reduce dependance on foreign oil and greenhouse gas emissions. If we could run our vehicles on 100% ethanol, the difference would be noticeable.

Contributes to Creation of Employment to the Country

When the use of ethanol fuel increases, it means more plantations of sugarcane, corn, and grains. It also means more ethanol fuel processing plants and that translates to job opportunities. Ethanol can also be branched out to produce alcoholic beverages leading to creation of job opportunities in the hospitality industry.

Opens up Untapped Agricultural Sector

The fact that ethanol fuel production relies mainly on agricultural produce, individuals will be shoved into the untapped agricultural sector, and this will uplift a countries economy. This act will guarantee ethanol fuel availability for many years. The need for increased production of corn and grains has set the farming industry booming.

Ethanol Fuel is a Source of Hydrogen

Although ethanol fuel is not perfect, researchers are working around the clock to beef up its efficiency to make it a reliable energy source by getting rid of its disadvantages. One disadvantage of ethanol fuel is that it has been reported to cause engine burns and corrosion. To be able to utilize it in a more productive way, researchers are looking to convert it into hydrogen form, which should uplift it as a formidable alternative source of fuel.

Variety of Sources of Raw Material

Although corn and sugarcane are the chief raw material for producing ethanol fuel, pretty much every crop or plant containing starch and sugar can be used.

Ethanol is Classifie   as a Renewable Energy Source

It's classified as a renewable resource because it's mainly as a consequence of conversion of energy from the sun into useful energy. The production of ethanol begins with the photosynthesis process, which enables sugarcane to thrive and later be processed into ethanol fuel.

Disadvantages of Ethanol Fuel

Requires Large Piece of Land

We've learned that ethanol is produced from corn, sugarcane, and grains. All these are crops that need to be grown in farms. For ethanol to meet the growing demand, it must be produced in large scale. This, essential, means that these very crops will have to be grown in large scale, which requires vast acres of land. The problem is that not everyone has that kind of land, so the only option is renting or leasing, which might add expenses to the budget. This aspect could also lead to destruction of natural habitats for most plants and animals.

Distillation Process is not Good for Environment

The process of distilling fermented corn or grain takes a long time and involves a lot of heat expenditure. The source of heat for distillation is mostly fossil fuel, and fossil fuels emit a lot of greenhouse gas, which is detrimental to the environment.

Spike in Food Prices

The chief ingredient in making ethanol is corn. If the demand for ethanol fuel skyrockets, the price of corn would also shoot up, and that would affect the cost of ethanol production. Other users of corn other than for fuel will also suffer, for example, those utilizing corn as an animal feed. Also, the lucrative prices of ethanol fuel could trigger most farmers to abandon food crops for ethanol production, which might also lead to an increase in food prices.

Water Attraction

Pure ethanol has high affinity for water, and it's able to absorb any trace around it or from the atmosphere. This fact is also true for those blends of gasoline and ethanol used to power vehicles. The fact that ethanol has high water attraction capabilities means that it's difficult to obtain it in its purest form since there will somehow be a trace of water. In fact, manufacturers normally indicate 99.8% pure ethanol. This is especially dangerous for marine users than regular road car users.

When water finds way into a storage or fuel tank, it goes to the bottom of tank since water is denser than fuel. This will lead to a plethora of small and big engine problems for your vehicle. The water attraction property of ethanol is the reason why it's transported by railroad or auto transport.

Difficul to Vaporize

Pure ethanol is hard to vaporize. This makes starting a car in cold conditions almost difficult, which is why a number of vehicle owners make a point to retain a little petrol, for instance, E85 cars that use 15% petroleum and 85% ethanol.

A common blend used these days is E85 i.e. 85% Ethanol and 15% gasoline. The mileage provided by this blend is lesser than that of pure gasoline or the E10 (10% Ethanol) blend. However, the benefit of using the E85 blend is that the oil remains clean for a longer time, there is lesser stress on the engine and the overall engine maintenance reduces. The cost of lower mileage gets covered up thanks to these small benefits. Not to mention, the overall reduction of your carbon footprint, which is the one benefit from the use of Ethanol fuel that everybody should aspire for.

Cellulosic Ethanol

Cellulosic ethanol is ethanol (ethyl alcohol) produced from cellulose (the stringy fiber of a plant) rather than from the plant's seeds or fruit. It is a biofuel produced from grasses, wood, algae, or other plants. The fibrous parts of the plants are mostly inedible to animals, including humans, except for ruminants (grazing, cud-chewing animals such as cows or sheep).

Considerable interest in cellulosic ethanol exists due to its important economical potential. Growth of cellulose by plants is a mechanism that captures and stores solar energy chemically in nontoxic ways with resultant supplies that are easy to transport and store. Additionally, transport may be unneeded anyway, because grasses or trees can grow almost anywhere temperate. This is why commercially practical cellulosic ethanol is widely viewed as a next level of development for the biofuel industry that could reduce demand for oil and gas drilling and even nuclear power in ways that grain-based ethanol fuel alone cannot. Potential exists for the many benefits of carbonaceous liquid fuels and petrochemicals (which today's standard of living depends on) but in a carbon cycle–balanced and renewable way (recycling surface and atmosphere carbon instead of pumping underground carbon up into it and thus adding to it). Commercially practical cellulosic alcohol could also avoid one of the problems with today's conventional (grain-based) biofuels, which is that they set up competition for grain with food purposes, potentially driving up the price of food. To date, what stands in the way of these goals is that production of cellulosic alcohol is not yet sufficiently practical on a commercial scale.

Cellulosic ethanol is a type of biofuel produced from lignocellulose, a structural material that comprises much of the mass of plants. Lignocellulose is composed mainly of cellulose, hemicellulose and lignin. Corn stover, *Panicum virgatum* (switchgrass), *Miscanthus* grass species, wood chips and the byproducts of lawn and tree maintenance are some of the more popular cellulosic materials for ethanol production. Production of ethanol from lignocellulose has the advantage of abundant and diverse raw material compared to sources such as corn and cane sugars, but requires a greater amount of processing to make the sugar monomers available to the microorganisms typically used to produce ethanol by fermentation.

Switchgrass and *Miscanthus* are the major biomass materials being studied today, due to their high productivity per acre. Cellulose, however, is contained in nearly every natural, free-growing plant, tree, and bush, in meadows, forests, and fields all over the world without agricultural effort or cost needed to make it grow.

One of the benefits of cellulosic ethanol is it reduces greenhouse gas emissions (GHG) by 85% over reformulated gasoline. By contrast, starch ethanol (e.g., from corn), which most frequently uses natural gas to provide energy for the process, may not reduce GHG emissions at all depending on how the starch-based feedstock is produced. According to the National Academy of Sciences in 2011, there is no commercially viable bio-refinery in existence to convert lignocellulosic biomass to fuel. Absence of production of cellulosic ethanol in the quantities required by the regulation was the basis of a United States Court of Appeals for the District of Columbia decision announced January 25, 2013, voiding a requirement imposed on car and truck fuel producers in the United States by the Environmental Protection Agency requiring addition of cellulosic biofuels to their products. These issues, along with many other difficult production challenges, led George Washington University policy researchers to state that "in the short term, [cellulosic] ethanol cannot meet the energy security and environmental goals of a gasoline alternative."

Production Methods

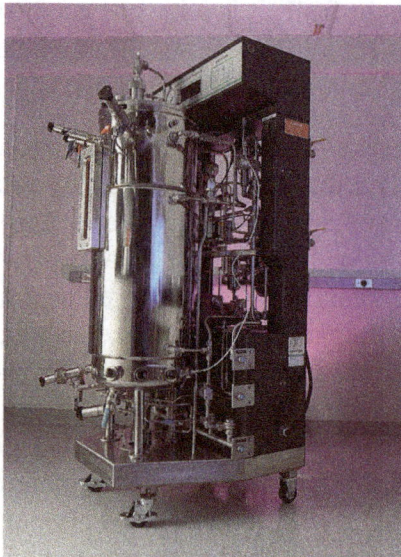

Bioreactor for cellulosic ethanol research.

The two ways of producing ethanol from cellulose are:

- Cellulolysis processes which consist of hydrolysis on pretreated lignocellulosic materials, using enzymes to break complex cellulose into simple sugars such as glucose, followed by fermentation and distillation.

- Gasification that transforms the lignocellulosic raw material into gaseous carbon monoxide and hydrogen. These gases can be converted to ethanol by fermentation or chemical catalysis.

As is normal for pure ethanol production, these methods include distillation.

Cellulolysis Biological Approach

The stages to produce ethanol using a biological approach are:

1. A "pretreatment" phase, to make the lignocellulosic material such as wood or straw amenable to hydrolysis.

2. Cellulose hydrolysis (that is, cellulolysis) with cellulases, to break down the molecules into sugars.

3. Separation of the sugar solution from the residual materials, notably lignin.

4. Microbial fermentation of the sugar solution.

5. Distillation to produce roughly 95% pure alcohol.

6. Dehydration by molecular sieves to bring the ethanol concentration to over 99.5%.

In 2010, a genetically engineered yeast strain was developed to produce its own cellulose-digesting enzymes. Assuming this technology can be scaled to industrial levels, it would eliminate one or more steps of cellulolysis, reducing both the time required and costs of production.

Although lignocellulose is the most abundant plant material resource, its usability is curtailed by its rigid structure. As a result, an effective pretreatment is needed to liberate the cellulose from the lignin seal and its crystalline structure so as to render it accessible for a subsequent hydrolysis step. By far, most pretreatments are done through physical or chemical means. To achieve higher efficiency, both physical and chemical pretreatments are required. Physical pretreatment is often called size reduction to reduce biomass physical size. Chemical pretreatment is to remove chemical barriers so the enzymes can have access to cellulose for microbial reactions.

To date, the available pretreatment techniques include acid hydrolysis, steam explosion, ammonia fiber expansion, organosolv, sulfite pretreatment, AVAP® (SO2-ethanol-water) fractionation, alkaline wet oxidation and ozone pretreatment. Besides effective cellulose liberation, an ideal pretreatment has to minimize the formation of degradation products because of their inhibitory effects on subsequent hydrolysis and fermentation processes. The presence of inhibitors will not only further complicate the ethanol production but also increase the cost of production due to entailed detoxification steps. Even though pretreatment by acid hydrolysis is probably the oldest and most studied pretreatment technique, it produces several potent inhibitors including furfural and hydroxymethyl furfural (HMF) which are by far regarded as the most toxic inhibitors present in lignocellulosic hydrolysate. Ammonia Fiber Expansion (AFEX) is a promising pretreatment with no inhibitory effect in resulting hydrolysate.

Most pretreatment processes are not effective when applied to feedstocks with high lignin content, such as forest biomass. Organosolv, SPORL ('sulfite pretreatment to overcome recalcitrance of lignocellulose') and SO2-ethanol-water (AVAP®) processes are the three processes that can achieve over 90% cellulose conversion for forest biomass, especially those of softwood species. SPORL is the most energy efficient (sugar production per unit energy consumption in pretreatment) and robust process for pretreatment of forest biomass with very low production of fermentation inhibitors. Organosolv pulping is particularly effective for hardwoods and offers easy recovery of a hydrophobic lignin product by dilution and precipitation. AVAP® process effectively fractionates all types of lignocellulosics into clean highly digestible cellulose, undegraded hemicellulose sugars, reactive lignin and lignosulfonates, and is characterized by efficient recovery of chemicals.

There are two major cellulose hydrolysis (cellulolysis) processes: a chemical reaction using acids, or an enzymatic reaction use cellulases.

Cellulolytic Processes

The cellulose molecules are composed of long chains of sugar molecules. In the hydrolysis of cellulose (that is, cellulolysis), these chains are broken down to free the sugar before it is fermented for alcohol production.

Chemical Hydrolysis

In the traditional methods developed in the 19th century and at the beginning of the 20th century, hydrolysis is performed by attacking the cellulose with an acid. Dilute acid may be used under high heat and high pressure, or more concentrated acid can be used at lower temperatures and atmospheric pressure. A decrystalized cellulosic mixture of acid and sugars reacts in the presence of water to complete individual sugar molecules (hydrolysis). The product from this hydrolysis is then neutralized and yeast fermentation is used to produce ethanol. As mentioned, a significant obstacle to the dilute acid process is that the hydrolysis is so harsh that toxic degradation products are produced that can interfere with fermentation. BlueFire Renewables uses concentrated acid because it does not produce nearly as many fermentation inhibitors, but must be separated from the sugar stream for recycle [simulated moving bed (SMB) chromatographic separation, for example] to be commercially attractive.

Agricultural Research Service scientists found they can access and ferment almost all of the remaining sugars in wheat straw. The sugars are located in the plant's cell walls, which are notoriously difficult to break down. To access these sugars, scientists pretreated the wheat straw with alkaline peroxide, and then used specialized enzymes to break down the cell walls. This method produced 93 US gallons (350 L) of ethanol per ton of wheat straw.

Enzymatic Hydrolysis

Cellulose chains can be broken into glucose molecules by cellulase enzymes. This reaction occurs at body temperature in the stomachs of ruminants such as cattle and sheep, where the enzymes are produced by microbes. This process uses several enzymes at various stages of this conversion. Using a similar enzymatic system, lignocellulosic materials can be enzymatically hydrolyzed at a relatively mild condition (50 °C and pH 5), thus enabling effective cellulose breakdown without the formation of byproducts that would otherwise inhibit enzyme activity. All major pretreatment methods, including dilute acid, require an enzymatic hydrolysis step to achieve high sugar yield for ethanol fermentation. Currently, most pretreatment studies have been laboratory-based, but companies are exploring means to transition from the laboratory to pilot, or production scale.

Various enzyme companies have also contributed significant technological breakthroughs in cellulosic ethanol through the mass production of enzymes for hydrolysis at competitive prices.

The fungus *Trichoderma reesei* is used by Iogen Corporation to secrete "specially engineered enzymes" for an enzymatic hydrolysis process. Their raw material (wood or straw) has to be pre-treated to make it amenable to hydrolysis.

Another Canadian company, SunOpta, uses steam explosion pretreatment, providing its technology to Verenium (formerly Celunol Corporation)'s facility in Jennings, Louisiana, Abengoa's facility in Salamanca, Spain, and a China Resources Alcohol Corporation in Zhaodong. The CRAC production facility uses corn stover as raw material.

Genencor and Novozymes have received United States Department of Energy funding for research into reducing the cost of cellulases, key enzymes in the production of cellulosic ethanol by enzymatic hydrolysis. A recent breakthrough in this regard was the discovery and inclusion of *lytic polysaccharide monooxygenases*. These enzymes are capable of boosting significantly the action of other cellulases by oxidatively attacking a polysaccharide substrate.

Other enzyme companies, such as Dyadic International, are developing genetically engineered fungi which would produce large volumes of cellulase, xylanase and hemicellulase enzymes, which can be used to convert agricultural residues such as corn stover, distiller grains, wheat straw and sugarcane bagasse and energy crops such as switchgrass into fermentable sugars which may be used to produce cellulosic ethanol.

In 2010, BP Biofuels bought out the cellulosic ethanol venture share of Verenium, which had itself been formed by the merger of Diversa and Celunol, and with which it jointly owned and operated a 1.4-million-US-gallon (5,300 m³) per year demonstration plant in Jennings, LA, and the laboratory facilities and staff in San Diego, CA. BP Biofuels continues to operate these facilities, and has begun first phases to construct commercial facilities. Ethanol produced in the Jennings facility was shipped to London and blended with gasoline to provide fuel for the Olympics.

KL Energy Corporation, formerly KL Process Design Group, began commercial operation of a 1.5-million-US-gallon (5,700 m³) per year cellulosic ethanol facility in Upton, WY in the last quarter of 2007. The Western Biomass Energy facility is currently achieving yields of 40–45 US gallons (150–170 L) per dry ton. It is the first operating commercial cellulosic ethanol facility in the nation. The KL Energy process uses a thermomechanical breakdown and enzymatic conversion process. The primary feedstock is soft wood, but lab tests have already proven the KL Energy process on wine pomace, sugarcane bagasse, municipal solid waste, and switchgrass.

Microbial Fermentation

Traditionally, baker's yeast (*Saccharomyces cerevisiae*), has long been used in the brewery industry to produce ethanol from hexoses (six-carbon sugars). Due to the complex nature of the carbohydrates present in lignocellulosic biomass, a significant amount of xylose and arabinose (five-carbon sugars derived from the hemicellulose portion of the lignocellulose) is also present in the hydrolysate. For example, in the hydrolysate of corn stover, approximately 30% of the total fermentable sugars is xylose. As a result, the ability of the fermenting microorganisms to use the whole range of sugars available from the hydrolysate is vital to increase the economic competitiveness of cellulosic ethanol and potentially biobased proteins.

In recent years, metabolic engineering for microorganisms used in fuel ethanol production has shown significant progress. Besides *Saccharomyces cerevisiae*, microorganisms such as *Zymomonas mobilis* and *Escherichia coli* have been targeted through metabolic engineering for cellulosic ethanol production. An attraction towards alternative fermentation organism is its ability to ferment five carbon sugars improving the yield of the feed stock. This ability is often found in bacteria based organisms.

Recently, engineered yeasts have been described efficiently fermenting xylose, and arabinose, and even both together. Yeast cells are especially attractive for cellulosic ethanol processes because they have been used in biotechnology for hundreds of years, are tolerant to high ethanol and inhibitor concentrations and can grow at low pH values to reduce bacterial contamination.

Combined Hydrolysis and Fermentation

Some species of bacteria have been found capable of direct conversion of a cellulose substrate into ethanol. One example is *Clostridium thermocellum*, which uses a complex cellulosome to break down cellulose and synthesize ethanol. However, *C. thermocellum* also produces other products during cellulose metabolism, including acetate and lactate, in addition to ethanol, lowering the efficiency of the process. Some research efforts are directed to optimizing ethanol production by genetically engineering bacteria that focus on the ethanol-producing pathway.

Gasificatio Process (Thermochemical Approach)

Fluidized Bed Gasifier.

The gasification process does not rely on chemical decomposition of the cellulose chain (cellulolysis). Instead of breaking the cellulose into sugar molecules, the carbon in the raw material is converted into synthesis gas, using what amounts to partial combustion. The carbon monoxide, carbon dioxide and hydrogen may then be fed into a special kind of fermenter. Instead of sugar fermentation with yeast, this process uses *Clostridium ljungdahlii* bacteria. This microorganism will ingest carbon monoxide, carbon dioxide and hydrogen and produce ethanol and water. The process can thus be broken into three steps:

1. Gasification — Complex carbon-based molecules are broken apart to access the carbon as carbon monoxide, carbon dioxide and hydrogen.

2. Fermentation — Convert the carbon monoxide, carbon dioxide and hydrogen into ethanol using the *Clostridium ljungdahlii* organism.

3. Distillation — Ethanol is separated from water.

A recent study has found another *Clostridium* bacterium that seems to be twice as efficient in making ethanol from carbon monoxide as the one mentioned above.

Alternatively, the synthesis gas from gasification may be fed to a catalytic reactor where it is used to produce ethanol and other higher alcohols through a thermochemical process. This process can also generate other types of liquid fuels, an alternative concept successfully demonstrated by the Montreal-based company Enerkem at their facility in Westbury, Quebec.

Hemicellulose to Ethanol

Studies are intensively conducted to develop economic methods to convert both cellulose and hemicellulose to ethanol. Fermentation of glucose, the main product of cellulose hydrolyzate, to ethanol is an already established and efficient technique. However, conversion of xylose, the pentose sugar of hemicellulose hydrolyzate, is a limiting factor, especially in the presence of glucose. Moreover, it cannot be disregarded as hemicellulose will increase the efficiency and cost-effectiveness of cellulosic ethanol production.

Sakamoto et al. show the potential of genetic engineering microbes to express hemicellulase enzymes. The researchers created a recombinant Saccharomyces cerevisiae strain that was able to:

1. Hydrolyze hemicellulase through codisplaying endoxylanase on its cell surface.

2. Assimilate xylose by expression of xylose reductase and xylitol dehydrogenase.

The strain was able to convert rice straw hydrolyzate to ethanol, which contains hemicellulosic components. Moreover, it was able to produce 2.5x more ethanol than the control strain, showing the highly effective process of cell surface-engineering to produce ethanol.

Economics

The shift to a renewable fuel resource has been a target for many years now. However, most of its production is with the use of corn ethanol. In the year 2000, only 6.2 billion liters were produced in the United States, but this number has expanded over 800% to 50 billion litres in just a decade (2010). Government pressures to shift to renewable fuel resources have been apparent since the U.S. Environmental Protection Agency implemented the 2007 Renewable Fuel Standard (RFS), which required that a certain percentage of renewable fuel be included in fuel products. The shift to cellulosic ethanol production from corn ethanol has been strongly promoted by the US government. Even with these policies in place and the government's attempts to create a market for cellulose ethanol, there was no commercial production of this fuel in 2010 and 2011. The Energy Independence and Security Act originally set goals of 100 million, 250 million, and 500 million gallons for the years 2010, 2011, and 2012 respectively. However, as of 2012 it was projected that the production of cellulosic ethanol would be approximately 10.5 million gallons--far from its target. In 2007 alone, the US government provided 1 billion US dollars for cellulosic ethanol projects, while China invested 500 million US dollars into cellulosic ethanol research.

Due to lack of existing commercialized plant data, it is difficult to determine the exact method of production that will be most commonly employed. Model systems try to compare technologies and costs, but these models cannot be applied to commercial-plant costs. Currently, there are many pilot and demonstration facilities open that exhibit cellulosic production on a smaller scale.

Start-up costs for pilot scale lignocellulosic ethanol plants are high. On 28 February 2007, the U.S. Department of Energy announced $385 million in grant funding to six cellulosic ethanol plants. This grant funding accounts for 40% of the investment costs. The remaining 60% comes from the promoters of those facilities. Hence, a total of $1 billion will be invested for approximately 140-million-US-gallon (530,000 m³) capacity. This translates into $7/annual gallon production capacity in capital investment costs for pilot plants; future capital costs are expected to be lower. Corn-to-ethanol plants cost roughly $1–3/annual gallon capacity, though the cost of the corn itself is considerably greater than for switchgrass or waste biomass.

As of 2007, ethanol is produced mostly from sugars or starches, obtained from fruits and grains. In contrast, cellulosic ethanol is obtained from cellulose, the main component of wood, straw, and much of the structure of plants. Since cellulose cannot be digested by humans, the production of cellulose does not compete with the production of food, other than conversion of land from food production to cellulose production (which has recently started to become an issue, due to rising wheat prices.) The price per ton of the raw material is thus much cheaper than that of grains or fruits. Moreover, since cellulose is the main component of plants, the whole plant can be harvested. This results in much better yields—up to 10 short tons per acre (22 t/ha), instead of 4-5 short tons/acre (9–11 t/ha) for the best crops of grain.

The raw material is plentiful. An estimated 323 million tons of cellulose-containing raw materials which could be used to create ethanol are thrown away each year in US alone. This includes 36.8 million dry tons of urban wood wastes, 90.5 million dry tons of primary mill residues, 45 million dry tons of forest residues, and 150.7 million dry tons of corn stover and wheat straw. Transforming them into ethanol using efficient and cost-effective hemi(cellulase) enzymes or other processes might provide as much as 30% of the current fuel consumption in the United States. Moreover, even land marginal for agriculture could be planted with cellulose-producing crops, such as switchgrass, resulting in enough production to substitute for all the current oil imports into the United States.

Paper, cardboard, and packaging comprise a substantial part of the solid waste sent to landfills in the United States each day, 41.26% of all organic municipal solid waste (MSW) according to California Integrated Waste Management Board's city profiles. These city profiles account for accumulation of 612.3 short tons (555.5 t) daily per landfill where an average population density of 2,413 per square mile persists. All these, except gypsum board, contain cellulose, which is transformable into cellulosic ethanol. This may have additional environmental benefits because decomposition of these products produces methane, a potent greenhouse gas.

Reduction of the disposal of solid waste through cellulosic ethanol conversion would reduce solid waste disposal costs by local and state governments. It is estimated that each person in the US throws away 4.4 lb (2.0 kg) of trash each day, of which 37% contains waste paper, which is largely cellulose. That computes to 244 thousand tons per day of discarded waste paper that contains cellulose. The raw material to produce cellulosic ethanol is not only free, it has a negative cost—i.e., ethanol producers can get paid to take it away.

In June 2006, a U.S. Senate hearing was told the current cost of producing cellulosic ethanol is

US$2.25 per US gallon (US$0.59/litre), primarily due to the current poor conversion efficiency. At that price, it would cost about $120 to substitute a barrel of oil (42 US gallons (160 L)), taking into account the lower energy content of ethanol. However, the Department of Energy is optimistic and has requested a doubling of research funding. The same Senate hearing was told the research target was to reduce the cost of production to US$1.07 per US gallon (US$0.28/litre) by 2012. "The production of cellulosic ethanol represents not only a step toward true energy diversity for the country, but a very cost-effective alternative to fossil fuels. It is advanced weaponry in the war on oil," said Vinod Khosla, managing partner of Khosla Ventures, who recently told a Reuters Global Biofuels Summit that he could see cellulosic fuel prices sinking to $1 per gallon within ten years.

In September 2010, a report by Bloomberg analyzed the European biomass infrastructure and future refinery development. Estimated prices for a litre of ethanol in August 2010 are EUR 0.51 for 1 g and 0.71 for 2 g. The report suggested Europe should copy the current US subsidies of up to $50 per dry tonne.

Recently on October 25, 2012, BP, one of the leaders in fuel products, announced the cancellation of their proposed $350 million commercial-scale plant. It was estimated that the plant would be producing 36 million gallons a year at its location in Highlands County of Florida. BP has still provided 500 million US dollars for biofuel research at the Energy Biosciences Institute. General Motors (GM) has also invested into cellulosic companies more specifically Mascoma and Coskata. There are many other companies in construction or heading towards it. Abengoa is building a 25 million-gallon per year plant in \ technology platform based on the fungus Myceliophthora thermophila to convert lignocellulose into fermentable sugars. Poet is also in midst of producing a 200 million dollar, 25-million-gallon per year in Emmetsburg, Iowa. Mascoma now partnered with Valero has declared their intention to build a 20 million gallon per year in Kinross, Michigan. China Alcohol Resource Corporation has developed a 6.4 million liter cellulosic ethanol plant under continuous operation.

Also, since 2013, the Brazilian company GranBio is working to become a producer of biofuels and biochemicals. The family-held company is commissioning an 82 million liters per year (22 MMgy) cellulosic ethanol plant (2G ethanol) in the state of Alagoas, Brazil, which will be the first industrial facility of the group. GranBio's second generation ethanol facility is integrated to a first generation ethanol plant operated by Grupo Carlos Lyra, uses process technology from Beta Renewables, enzymes from Novozymes and yeast from DSM. Breaking ground in January 2013, the plant is in final commissioning. According to GranBio Annual Financial Records, the total investment was 208 million US Dollars.

Enzyme-cost Barrier

Cellulases and hemicellulases used in the production of cellulosic ethanol are more expensive compared to their first generation counterparts. Enzymes required for maize grain ethanol production cost 2.64-5.28 US dollars per cubic meter of ethanol produced. Enzymes for cellulosic ethanol production are projected to cost 79.25 US dollars, meaning they are 20-40 times more expensive. The cost differences are attributed to quantity required. The cellulase family of enzymes have a one to two order smaller magnitude of efficiency. Therefore, it requires 40 to 100 times more of the enzyme to be present in its production. For each ton of biomass it requires 15-25 kilograms of enzyme. More recent estimates are lower, suggesting 1 kg of enzyme per dry tonne of biomass feedstock. There is also relatively high capital costs associated with the long incubation times for the vessel that perform enzymatic hydrolysis. Altogether, enzymes comprise a significant portion of 20-40% for cellulosic ethanol production. A recent paper estimates the range at 13-36% of cash costs, with a key factor being how the cellulase enzyme is produced. For cellulase produced offsite, enzyme production amounts to 36% of cash cost. For enzyme produced onsite in a separate plant, the fraction is 29%; for integrated enzyme production, the faction is 13%. One of the key benefits of integrated production is that biomass instead of glucose is the enzyme growth medium. Biomass costs less, and it makes the resulting cellulosic ethanol a 100% second-generation biofuel, i.e., it uses no 'food for fuel'.

Feedstocks

In general there are two types of feedstocks: forest (woody) Biomass and agricultural biomass. In the US, about 1.4 billion dry tons of biomass can be sustainably produced annually. About 370 million tons or 30% are forest biomass. Forest biomass has higher cellulose and lignin content and lower hemicellulose and ash content than agricultural biomass. Because of the difficulties and low ethanol yield in fermenting pretreatment hydrolysate, especially those with very high 5 carbon hemicellulose sugars such as xylose, forest biomass has significant advantages over agricultural biomass. Forest biomass also has high density which significantly reduces transportation cost. It can be harvested year around which eliminates long term storage. The close to zero ash content of forest biomass significantly reduces dead load in transportation and processing. To meet the needs for biodiversity, forest biomass will be an important biomass feedstock supply mix in the future biobased economy. However, forest biomass is much more recalcitrant than agricultural biomass. Recently, the USDA Forest Products Laboratory together with the University of Wisconsin–Madison developed efficient technologies that can overcome the strong recalcitrance of forest (woody) biomass including those of softwood species that have low xylan content. Short-rotation intensive culture or tree farming can offer an almost unlimited opportunity for forest biomass production.

Woodchips from slashes and tree tops and saw dust from saw mills, and waste paper pulp are common forest biomass feedstocks for cellulosic ethanol production.

The following are a few examples of agricultural biomass:

Switchgrass (Pan*icum virgatum*) is a native tallgrass prairie grass. Known for its hardiness and rapid growth, this perennial grows during the warm months to heights of 2–6 feet. Switchgrass can be grown in most parts of the United States, including swamplands, plains, streams, and along the shores & *interstate highways*. It is *self-seeding* (no tractor for sowing, only for mowing), resistant to many diseases and pests, & can produce high yields with low applications of fertilizer and other chemicals. It is also tolerant to poor soils, flooding, & drought; improves soil quality and prevents erosion due its type of root system.

Switchgrass is an approved cover crop for land protected under the federal Conservation Reserve Program (CRP). CRP is a government program that pays producers a fee for not growing crops on land on which crops recently grew. This program reduces soil erosion, enhances water quality, and increases wildlife habitat. CRP land serves as a habitat for upland game, such as pheasants and ducks, and a number of insects. Switchgrass for biofuel production has been considered for use on Conservation Reserve Program (CRP) land, which could increase ecological sustainability and lower the cost of the CRP program. However, CRP rules would have to be modified to allow this economic use of the CRP land.

Miscanthus × giganteus is another viable feedstock for cellulosic ethanol production. This species of grass is native to Asia and is the sterile triploid hybrid of *Miscanthus sinensis* and *Miscanthus sacchariflorus*. It can grow up to 12 feet (3.7 m) tall with little water or fertilizer input. Miscanthus is similar to switchgrass with respect to cold and drought tolerance and water use efficiency. Miscanthus is commercially grown in the European Union as a combustible energy source.

Corn cobs and corn stover are the most popular agricultural biomass.

It has been suggested that Kudzu may become a valuable source of biomass.

Environmental Effects

The environmental impact from the production of fuels is an important factor in determining its feasibility as an alternative to fossil fuels. Over the long run, small differences in production cost, environmental ramifications, and energy output may have large effects. It has been found that cellulosic ethanol can produce a positive net energy output. The reduction in green house gas (GHG) emissions from corn ethanol and cellulosic ethanol compared with fossil fuels is drastic. Corn ethanol may reduce overall GHG emissions by about 13%, while that figure is around 88% or greater for cellulosic ethanol. As well, cellulosic ethanol can reduce carbon dioxide emissions to nearly zero.

Croplands

A major concern for the viability of current alternative fuels is the cropland needed to produce the required materials. For example, the production of corn for corn ethanol fuel competes with cropland that may be used for food growth and other feedstocks. The difference between this and cellulosic ethanol production is that cellulosic material is widely available and is derived from a large resource of things. Some crops used for cellulosic ethanol production include switchgrass, corn stover, and hybrid poplar. These crops are fast-growing and can be grown on many types of land which makes them more versatile. Cellulosic ethanol can also be made from wood residues (chips and sawdust), municipal solid waste such as trash or garbage, paper and sewage sludge, cereal straws and grasses. It is particularly the non-edible portions of plant material which are used to make cellulosic ethanol, which also minimizes the potential cost of using food products in production.

The effectiveness of growing crops for the purpose of biomass can vary tremendously depending on the geographical location of the plot. For example, factors such as precipitation and sunlight exposure may greatly effect the energy input required to maintain the crops, and therefore effect the overall energy output. A study done over five years showed that growing and managing switchgrass exclusively as a biomass energy crop can produce 500% or more renewable energy than is consumed during production. The levels of GHG emissions and carbon dioxide were also drastically decreased from using cellulosic ethanol compared with traditional gasoline.

Corn-based vs. Grass-based

Summary of Searchinger et al. comparison of corn ethanol and gasoline GHG emissions with and without land use change (Grams of CO_2 released per megajoule of energy in fuel)				
Fuel type (U.S.)	Carbon intensity	Reduction GHG	Carbonintensity+ ILUC	Reduction GHG
Gasoline	92	-	92	-
Corn ethanol	74	-20%	177	+93%
Cellulosic ethanol	28	-70%	138	+50%

In 2008, there was only a small amount of switchgrass dedicated for ethanol production. In order for it to be grown on a large-scale production it must compete with existing uses of agricultural land, mainly for the production of crop commodities. Of the United States' 2.26 billion acres (9.1 million km²) of unsubmerged land, 33% are forestland, 26% pastureland and grassland, and 20% crop land. A study done by the U.S. Departments of Energy and Agriculture in 2005 determined whether there were enough available land resources to sustain production of over 1 billion dry tons of biomass annually to replace 30% or more of the nation's current use of liquid transportation fuels. The study found that there could be 1.3 billion dry tons of biomass available for ethanol use, by making little changes in agricultural and forestry practices and meeting the demands for forestry products, food, and fiber. A recent study done by the University of Tennessee reported that as many as 100 million acres (400,000 km², or 154,000 sq mi) of cropland and pasture will need to be allocated to switchgrass production in order to offset petroleum use by 25 percent.

Currently, corn is easier and less expensive to process into ethanol in comparison to cellulosic ethanol. The Department of Energy estimates that it costs about $2.20 per gallon to produce cellulosic ethanol, which is twice as much as ethanol from corn. Enzymes that destroy plant cell wall tissue cost 30 to 50 cents per gallon of ethanol compared to 3 cents per gallon for corn. The Department of Energy hopes to reduce production cost to $1.07 per gallon by 2012 to be effective. However, cellulosic biomass is cheaper to produce than corn, because it requires fewer inputs, such as energy, fertilizer, herbicide, and is accompanied by less soil erosion and improved soil fertility. Additionally, nonfermentable and unconverted solids left after making ethanol can be burned to provide the fuel needed to operate the conversion plant and produce electricity. Energy used to run corn-based ethanol plants is derived from coal and natural gas. The Institute for Local Self-Reliance estimates the cost of cellulosic ethanol from the first generation of commercial plants will be in the $1.90–$2.25 per gallon range, excluding incentives. This compares to the current cost of $1.20–$1.50 per gallon for ethanol from corn and the current retail price of over $4.00 per gallon for regular gasoline (which is subsidized and taxed).

One of the major reasons for increasing the use of biofuels is to reduce greenhouse gas emissions. In comparison to gasoline, ethanol burns cleaner, thus putting less carbon dioxide and overall pollution in the air. Additionally, only low levels of smog are produced from combustion. According to the U.S. Department of Energy, ethanol from cellulose reduces greenhouse gas emission by 86 percent when compared to gasoline and to corn-based ethanol, which decreases emissions by 52 percent. Carbon dioxide gas emissions are shown to be 85% lower than those from gasoline. Cellulosic ethanol contributes little to the greenhouse effect and has a five times better net energy balance than corn-based ethanol. When used as a fuel, cellulosic ethanol releases less sulfur, carbon monoxide, particulates, and greenhouse gases. Cellulosic ethanol should earn producers carbon reduction credits, higher than those given to producers who grow corn for ethanol, which is about 3 to 20 cents per gallon.

It takes 0.76 J of energy from fossil fuels to produce 1 J worth of ethanol from corn. This total includes the use of fossil fuels used for fertilizer, tractor fuel, ethanol plant operation, etc. Research has shown that fossil fuel can produce over five times the volume of ethanol from prairie grasses, according to Terry Riley, president of policy at the Theodore Roosevelt Conservation Partnership. The United States Department of Energy concludes that corn-based ethanol provides 26 percent more energy than it requires for production, while cellulosic ethanol provides 80 percent more energy. Cellulosic ethanol yields 80 percent more energy than is required to grow and convert it. The process of turning corn into ethanol requires about 1700 times (by volume) as much water as ethanol produced. Additionally, it leaves 12 times its volume in waste. Grain ethanol uses only the edible portion of the plant.

U.S. Environmental Protection Agency Draft life cycle GHG emissions reduction results for different time horizon and discount rate approaches (includes indirect land use change effects)		
Fuel Pathway	100 years + 2% discount rate	30 years +0% discount rate
Corn ethanol (natural gas dry mill)	-16%	+5%
Corn ethanol (Best case NG DM)	-39%	-18%
Corn ethanol (coal dry mill)	+13%	+34%
Corn ethanol (biomass dry mill)	-39%	-18%
Corn ethanol (biomass dry mill with combined heat and power)	-47%	-26%
Brazilian sugarcane ethanol	-44%	-26%
Cellulosic ethanol from switchgrass	-128%	-124%
Cellulosic ethanol from corn stover	-115%	-116%

Cellulose is not used for food and can be grown in all parts of the world. The entire plant can be used when producing cellulosic ethanol. Switchgrass yields twice as much ethanol per acre than corn. Therefore, less land is needed for production and thus less habitat fragmentation. Biomass materials require fewer inputs, such as fertilizer, herbicides, and other chemicals that can pose risks to wildlife. Their extensive roots improve soil quality, reduce erosion, and increase nutrient capture. Herbaceous energy crops reduce soil erosion by greater than 90%, when compared to conventional commodity crop production. This can translate into improved water quality for rural communities. Additionally, herbaceous energy crops add organic material to depleted soils and can increase soil carbon, which can have a direct effect on climate change, as soil carbon can absorb carbon dioxide in the air. As compared to commodity crop production, biomass reduces surface runoff and nitrogen transport. Switchgrass provides an environment for diverse wildlife habitation, mainly insects and ground birds. Conservation Reserve Program (CRP) land is composed of perennial grasses, which are used for cellulosic ethanol, and may be available for use.

For years American farmers have practiced row cropping, with crops such as sorghum and corn. Because of this, much is known about the effect of these practices on wildlife. The most significant effect of increased corn ethanol would be the additional land that would have to be converted to agricultural use and the increased erosion and fertilizer use that goes along with agricultural production. Increasing our ethanol production through the use of corn could produce negative effects on wildlife, the magnitude of which will depend on the scale of production and whether the land used for this increased production was formerly idle, in a natural state, or planted with other row crops. Another consideration is whether to plant a switchgrass monoculture or use a variety of grasses and other vegetation. While a mixture of vegetation types likely would provide better wildlife habitat, the technology has not yet developed to allow the processing of a mixture of different grass species or vegetation types into bioethanol. Of course, cellulosic ethanol production is still in its infancy, and the possibility of using diverse vegetation stands instead of monocultures deserves further exploration as research continues.

A study by Nobel Prize winner Paul Crutzen found ethanol produced from corn had a "net climate warming" effect when compared to oil when the full life cycle assessment properly considers the nitrous oxide (N20) emissions that occur during corn ethanol production. Crutzen found that crops with less nitrogen demand, such as grasses and woody coppice species, have more favourable climate impacts.

Cellulosic Ethanol Commercialization

Cellulosic ethanol commercialization is the process of building an industry out of methods of turning cellulose-containing organic matter into fuel. Companies such as Iogen, POET, and Abengoa are building refineries that can process biomass and turn it into ethanol, while companies such as DuPont, Diversa, Novozymes, and Dyadic are producing enzymes which could enable a cellulosic ethanol future. The shift from food crop feedstocks to waste residues and native grasses offers significant opportunities for a range of players, from farmers to biotechnology firms, and from project developers to investors.

The cellulosic ethanol industry developed some new commercial-scale plants in 2008. In the United States, plants totaling 12 million liters (3.17 million gal) per year were operational, and an additional 80 million liters (21.1 million gal.) per year of capacity - in 26 new plants - was under construction. In Canada, capacity of 6 million liters per year was operational. In Europe, several plants were operational in Germany, Spain, and Sweden, and capacity of 10 million liters per year was under construction.

Italy-based Mossi & Ghisolfi Group broke ground for its 13 MMgy cellulosic ethanol facility in northwestern Italy on April 12, 2011. The project will be the largest cellulosic ethanol project in the world, 10 times larger than any of the currently operating demonstration-scale facilities.

Xyleco An independent engineering firm conducted an ISO conformant comparative life cycle assessment (LCA) of Xyleco's patented process on a "cradle-to-grave" basis and concluded that the global warming potential of Xyleco ethanol is 83% lower than gasoline, 77% lower than corn ethanol and 40% lower than sugarcane ethanol.

Butanol

Butanol may be used as a fuel in an internal combustion engine. Because its longer hydrocarbon chain causes it to be fairly non-polar, it is more similar to gasoline than it is to ethanol. Butanol is a drop-in fuel and thus works in vehicles designed for use with gasoline without modification.

It has a four link hydrocarbon chain. It can be produced from biomass (as "biobutanol") as well as fossil fuels (as "petrobutanol"), but biobutanol and petrobutanol have the same chemical properties.

Production of Biobutanol

Butanol from biomass is called biobutanol. It can be used in unmodified gasoline engines. High cost of raw material is considered as one of the main barriers against commercial butanol fermentation. Using inexpensive and abundant feedstocks, e.g., corn stover, can enhance the process economic viability.

Technologies

Biobutanol can be produced by fermentation of biomass by the A.B.E. process. The process uses the bacterium *Clostridium acetobutylicum*, also known as the *Weizmann organism*, or *Clostridium beijerinckii*. It was Chaim Weizmann who first used *C. acetobutylicum* for the production of acetone from starch (with the main use of acetone being the making of Cordite) in 1916. The butanol was a by-product of fermentation (twice as much butanol was produced). The process also creates a recoverable amount of H_2 and a number of other by-products: acetic, lactic and propionic acids, isopropanol and ethanol.

Biobutanol can also be made using *Ralstonia eutropha* H16. This process requires the use of an electro-bioreactor and the input of carbon dioxide and electricity.

The difference from ethanol production is primarily in the fermentation of the feedstock and minor changes in distillation. The feedstocks are the same as for ethanol: energy crops such as sugar beets, sugar cane, corn grain, wheat and cassava, prospective non-food energy crops such as switchgrass and even guayule in North America, as well as agricultural byproducts such as bagasse, straw and corn stalks. According to DuPont, existing bioethanol plants can cost-effectively be retrofitted to biobutanol production.

Additionally, butanol production from biomass and agricultural byproducts could be more efficient (i.e. unit engine motive power delivered per unit solar energy consumed) than ethanol or methanol production.

Algae Butanol

Biobutanol can be made entirely with solar energy and nutrients, from algae (called Solalgal Fuel) or diatoms. Current yield is low.

Research

Although biofuel demand has risen to over one billion liters (about 260 million US gallons) yearly, fermentation remains a largely inefficient method of butanol production. Under normal living conditions, *Clostridium* bacterial communities have a low yield of butanol per gram of glucose. Obtaining higher yields of butanol involves manipulation of the metabolic networks within bacteria to prioritize the synthesis of the biofuel. Metabolic engineering and genetic engineering tools allow scientists to alter the states of reactions occurring in the organism, utilizing advanced techniques to create a bacterial strain capable of high butanol yield. Optimization can also be accomplished by the transfer of specific genetic information to other uni-cellular species, capitalizing on the traits of multiple organisms to achieve the highest rate of alcohol production.

Using Alternate Carbon Sources

One promising development in biobutanol production technology was discovered in the late summer of 2011—Tulane University's alternative fuel research scientists discovered a strain of *Clostridium*, called "TU-103", that can convert nearly any form of cellulose into butanol, and is the only known strain of *Clostridium*-genus bacteria that can do so in the presence of oxygen. The university's researchers have stated that the source of the "TU-103" *Clostridium* bacteria strain was most likely from the solid waste from one of the plains zebra at New Orleans' Audubon Zoo.

Metabolic engineering can be used to allow an organism to use a cheaper substrate such as glycerol instead of glucose. Because fermentation processes require glucose derived from foods, butanol production can negatively impact food supply. Glycerol is a good alternative source for butanol production. While glucose sources are valuable and limited, glycerol is abundant and has a low market price because it is a waste product of biodiesel production. Butanol production from glycerol is economically viable using metabolic pathways that exist in *Clostridium pasteurianum* bacterium.

A combination of succinate and ethanol can be fermented to produce butyrate (a precursor to butanol fuel) by utilizing the metabolic pathways present in a gram-positive anaerobic bacterium *Clostridium kluyveri*. Succinate is an intermediate of the TCA cycle, which metabolizes glucose. Anaerobic bacteria such as *Clostridium acetobutylicum* and *Clostridium saccharobutylicum* also contain these pathways. Succinate is first activated and then reduced by a two-step reaction to give 4-hydroxybutyrate, which is then metabolized further to crotonyl-coenzyme A (CoA) . Crotonyl-CoA is then converted to butyrate. The genes corresponding to these butanol production pathways from *Clostridium* were cloned to *E. coli*.

In 2012 researchers developed a method for storing electrical energy as chemical energy in higher alcohols (including butanol). These alcohols can then be used as liquid transportation fuels. The team led by James C. Liao genetically engineered lithoautotrophic microorganism known as *Ralstonia Eutropha* H16 to produce isobutanol and 3-methyl-1-butanol in an electro-bioreactor. Carbon dioxide is the sole carbon source for this process and electricity is used as the energetic component. The process they developed effectively separates the light and dark reactions that occur during photosynthesis. Solar panels are used to convert sunlight to electrical energy which is then converted using the microorganism to a chemical intermediate. The team is now in the process of scaling up the operation and believes this process will be more efficient than the biologic process.

Improving Efficienc

In late 2012, a new discovery made the alternative fuel butanol more attractive to the biofuel industry. Scientist Hao Feng found a method that could significantly reduce the cost of the energy involved in making butanol. His team was able to isolate the butanol molecules during the fermentation process so they do not kill the organisms, and produces 100% or more butanol. After the fermentation process, they used a process called cloud point separation to recover the butanol which used 4 times less energy.

Also in late 2012, using systems metabolic engineering, a Korean research team at the former Korea Advanced Institute of Science and Technology (KAIST) has succeeded in demonstrating an optimized process to increase butanol production by generating an engineered bacterium. Professor Sang Yup Lee at the Department of Chemical and Biomolecular Engineering, KAIST, Dr. Do Young Seung at GS Caltex, a large oil refining company in Korea, and Dr. Yu-Sin Jang at BioFuelChem, a startup butanol company in Korea, applied a systems metabolic engineering approach to improve the production of butanol through enhancing the performance of *Clostridium acetobutylicum*, one of the best known butanol-producing bacteria. In addition, the downstream process was optimized and an in situ recovery process was integrated to achieve higher butanol titer, yield, and productivity. The combination of systems metabolic engineering and bioprocess optimization resulted in the development of a process capable of producing more than 585 g of butanol from 1.8 kg of glucose, which allows the production of this important industrial solvent and advanced biofuel to be cost competitive.

The anaerobic bacteria *C. pasteurianum*, *C. acetobutylicum*, and other *Clostridium* species have metabolic pathways that convert glycerol to butanol through fermentation. However, the production of butanol from glycerol by fermentation in *C. Pasteurianum* is low. To counter this, a group of researchers used chemical mutagenesis to create a hyper butanol-producing strain. The best mutant strain in this study "MBEL_GLY2" produced 10.8 g of butanol per 80 g of glycerol fed to the bacteria. This improvement compares to the 7.6 g butanol produced by the native bacteria.

Many organisms have the capacity to produce butanol utilizing an acetyl-CoA dependent pathway. The main problem with this pathway is the first reaction involving the condensation of two acetyl-CoA molecules to acetoacetyl-CoA. This reaction is thermodynamically unfavorable due to the positive Gibbs free energy associated with it (dG = 6.8 kcal/mol). Some experimentation has been done that involves increasing the carbon storage through the organism by utilizing carbon dioxide flow through photosynthetic organisms. To follow in this path of research, scientists have attempted to engineer reaction pathways that can enable photosynthetic organisms (like blue-green algae) to produce butanol more efficiently.

A study done by Ethan I. Lan and James C. Liao attempted to utilize the ATP produced during photosynthesis in blue-green algae to work around the thermodynamically unfavorable acetyl-CoA condensation to acetoacetyl-CoA. The native system was re-engineered to have acetyl-CoA react with ATP and CO_2 to form an intermediate, malonyl-CoA. Malonyl-CoA then reacts with another acetyl-CoA to form the desired acetoacetyl-CoA. The energy release from ATP hydrolysis (dG = -7.3 kcal/mol) makes this pathway significantly more favorable than standard condensation. Because blue-green algae generate NADPH during photosynthesis, it can be assumed that the cofactor environment is NADPH rich. Therefore, the native reaction pathway was further engineered to use NADPH rather than the standard NADH. All of these adjustments led to a 4-fold increase in butanol production, showing the importance of ATP and cofactor driving forces as a design principle in pathway engineering.

Producers

DuPont and BP plan to make biobutanol the first product of their joint effort to develop, produce, and market next-generation biofuels. In Europe the Swiss company Butalco is developing genetically modified yeasts for the production of biobutanol from cellulosic materials. Gourmet Butanol, a United States-based company, is developing a process that utilizes fungi to convert organic waste into biobutanol.

Distribution

Butanol better tolerates water contamination and is less corrosive than ethanol and more suitable for distribution through existing pipelines for gasoline. In blends with diesel or gasoline, butanol is less likely to separate from this fuel than ethanol if the fuel is contaminated with water. There is also a vapor pressure co-blend synergy with butanol and gasoline containing ethanol, which facilitates ethanol blending. This facilitates storage and distribution of blended fuels.

Properties of Common Fuels

Fuel	Energy density	Air-fuel ratio	Specific energy	Heat of vaporization	RON	MON	AKI
Gasoline and biogasoline	32 MJ/L	14.7	2.9 MJ/kg air	0.36 MJ/kg	91–99	81–89	87-95
Butanol fuel	29.2 MJ/L	11.1	3.6 MJ/kg air	0.43 MJ/kg	96	78	87
Anhydrous Ethanol fuel	19.6 MJ/L	9.0	3.0 MJ/kg air	0.92 MJ/kg	107	89	
Methanol fuel	16 MJ/L	6.4	3.1 MJ/kg air	1.2 MJ/kg	106	92	

Energy Content and Effects on Fuel Economy

Switching a gasoline engine over to butanol would in theory result in a fuel consumption penalty of about 10% but butanol's effect on mileage is yet to be determined by a scientific study. While the energy density for any mixture of gasoline and butanol can be calculated, tests with other alcohol fuels have demonstrated that the effect on fuel economy is not proportional to the change in energy density.

Octane Rating

The octane rating of n-butanol is similar to that of gasoline but lower than that of ethanol and methanol. n-Butanol has a RON (Research Octane number) of 96 and a MON (Motor octane number) of 78 (with a resulting "(R+M)/2 pump octane number" of 87, as used in North America) while t-butanol has octane ratings of 105 RON and 89 MON. t-Butanol is used as an additive in gasoline but cannot be used as a fuel in its pure form because its relatively high melting point of 25.5 °C (79 °F) causes it to gel and solidify near room temperature. On the other hand, isobutanol has a lower melting point than n-butanol and favorable RON of 113 and MON of 94, and is thus much better suited to high fraction gasoline blends, blends with n-butanol, or as a standalone fuel.

A fuel with a higher octane rating is less prone to knocking (extremely rapid and spontaneous combustion by compression) and the control system of any modern car engine can take advantage of this by adjusting the ignition timing. This will improve energy efficiency, leading to a better fuel economy than the comparisons of energy content different fuels indicate. By increasing the compression ratio, further gains in fuel economy, power and torque can be achieved. Conversely, a fuel with lower octane rating is more prone to knocking and will lower efficiency. Knocking can also cause engine damage. Engines designed to run on 87 octane will not have any additional power/ fuel economy from being operated with higher octane fuel.

Air-fuel Ratio

Alcohol fuels, including butanol and ethanol, are partially oxidized and therefore need to run at richer mixtures than gasoline. Standard gasoline engines in cars can adjust the air-fuel ratio to accommodate variations in the fuel, but only within certain limits depending on model. If the limit is exceeded by running the engine on pure ethanol or a gasoline blend with a high percentage of ethanol, the engine will run lean, something which can critically damage components. Compared to ethanol, butanol can be mixed in higher ratios with gasoline for use in existing cars without the need for retrofit as the air-fuel ratio and energy content are closer to that of gasoline.

Specifi Energy

Alcohol fuels have less energy per unit weight and unit volume than gasoline. To make it possible to compare the net energy released per cycle a measure called the fuels specific energy is some-times used. It is defined as the energy released per air fuel ratio. The net energy released per cycle is higher for butanol than ethanol or methanol and about 10% higher than for gasoline.

Viscosity

The viscosity of alcohols increase with longer carbon chains. For this reason, butanol is used as an alternative to shorter alcohols when a more viscous solvent is desired. The kinematic viscosity of butanol is several times higher than that of gasoline and about as viscous as high quality diesel fuel.

Heat of Vaporization

The fuel in an engine has to be vaporized before it will burn. Insufficient vaporization is a known problem with alcohol fuels during cold starts in cold weather. As the heat of vaporization of bu-tanol is less than half of that of ethanol, an engine running on butanol should be easier to start in cold weather than one running on ethanol or methanol.

Potential Problems with the use of Butanol Fuel

The potential problems with the use of butanol are similar to those of ethanol:

- To match the combustion characteristics of gasoline, the utilization of butanol fuel as a substitute for gasoline requires fuel-flow increases (though butanol has only slightly less energy than gasoline, so the fuel-flow increase required is only minimal, maybe 10%, compared to 40% for ethanol).
- Alcohol-based fuels are not compatible with some fuel system components.
- Alcohol fuels may cause erroneous gas gauge readings in vehicles with capacitance fuel level gauging.
- While ethanol and methanol have lower energy densities than butanol, their higher octane number allows for greater compression ratio and efficiency.

- Butanol is one of many side products produced from current fermentation technologies; as a consequence, current fermentation technologies allow for very low yields of pure extracted butanol. When compared to ethanol, butanol is more fuel efficient as a fuel alternative, but ethanol can be produced at a much lower cost and with much greater yields.

- Butanol is toxic at a rate of 20g per liter and may need to undergo Tier 1 and Tier 2 health effects testing before being permitted as a primary fuel by the EPA.

Possible Butanol Fuel Mixtures

Standards for the blending of ethanol and methanol in gasoline exist in many countries, including the EU, the US and Brazil. Approximate equivalent butanol blends can be calculated from the relations between the stoichiometric fuel-air ratio of butanol, ethanol and gasoline. Common ethanol fuel mixtures for fuel sold as gasoline currently range from 5% to 10%. The share of butanol can be 60% greater than the equivalent ethanol share, which gives a range from 8% to 16%. "Equivalent" in this case refers only to the vehicle's ability to adjust to the fuel. Other properties such as energy density, viscosity and heat of vaporization will vary and may further limit the percentage of butanol that can be blended with gasoline. It was estimated that around 9.5 gigaliter (Gl) of gasoline can be saved and about 64.6 Gl of butanol-gasoline blend 16% (Bu16) can potentially be produced from corn residues in the US, which is equivalent to 11.8% of total domestic gasoline consumption.

Consumer acceptance may be limited due to the potentially offensive banana-like smell of n-butanol. Plans are underway to market a fuel that is 85% Ethanol and 15% Butanol (E85B), so existing E85 internal combustion engines can run on a 100% renewable fuel that could be made without using any fossil fuels. Because its longer hydrocarbon chain causes it to be fairly non-polar, it is more similar to gasoline than it is to ethanol. Butanol has been demonstrated to work in vehicles designed for use with gasoline without modification.

Algae Fuel

Algal biofuel is an alternative to fossil fuel, which is generated by specific algae species from carbon dioxide.

These algae species are primarily unicellular or diatom microalgae that produce high carbohydrate compositions suitable for ethanol production, high lipid compositions suitable for biodiesel production or high hydrocarbon compositions that are suitable for producing renewable distillates.

Increase in fuel costs and consumption, and depletion of natural fuel resources have created a demand for research into alternative forms of fuels in the last decade. Several companies and government agencies are funding research to try and make algae fuel production commercially viable.

The optimum selection of the algal species for biofuel production is based on the ability to sustain the culture, growth rate of the species, the biomass specific contents of proteins, carbohydrates, and lipids, and the overall supporting photosynthesis environment.

Fuels from Algae

The lipid (oily) part of the algae biomass can be extracted and converted into biodiesel by a process similar to that used for any other vegetable oil.

Butanol can be made from algae or diatoms using a solar-powered biorefinery. This fuel was found to have an energy density 10% less than gasoline, and greater than that of either methanol or ethanol.

The green waste left over from the algae oil extraction can be used to produce butanol.

Additionally, it was found that macroalgae can be fermented by Clostridria to form butanol and other solvents.

Biogasoline produced from algae biomass can be used in internal combustion engines. Methane, which is the chief component of natural gas, can be produced from algae using several methods - pyrolysis, gasification or anaerobic digestion.

Algae can also be used to produce green diesel, also known as renewable diesel through a hydro-cracking refinery process that breaks down molecules into shorter hydrocarbon chains used in diesel engines.

Algae Cultivation

Algae can produce up to 300 times more oil per unit area than conventional crops such as palms, soybeans, rapeseed or jatroba. The following three primary ways to grow algae for biofuel production have been identified:

Open Pond System

The open pond system is one of the easiest methods for the cultivation of algae with high-oil content. In this method, algae are grown in open ponds under very warm and sunny environments.

Although it is the simplest form of algae production, it also has some major drawbacks. Open systems using a monoculture are also vulnerable to viral infection. In order to enhance algae production using this method, water temperature needs to be controlled.

Closed-loop System

The closed-loop system was adapted to produce algae more quickly and efficiently than the open pond system. In this method, algae are placed in clear, plastic bags to allow them to be exposed to sunlight.

These bags are stacked high and protected from external elements using a cover. The clear plastic bag provides enough exposure to sunlight to increase the rate of algae production.

The greater the algae production, the greater the amount of oil will be extracted. Unlike the open pond method, this method prevents algal contamination.

Photobioreactors

Most of the companies that use algae as a source of biofuels employ borosilicate glass tubes known as bioreactors that are exposed to sunlight. Within these tubes, the algae can be grown at maximum levels, even to the point they can be harvested every day.

This method results in a very high output of algae and oil for producing biofuels. However, running a photobioreactor is more expensive and difficult than using the open pond system, but may provide a high level of control.

Benefit of Algal Fuel

The key benefits of algae and algal fuels are listed below:

- Algae require much less land to grow when compared to other traditional row crops, such as corn. In addition, algae can be grown on non-arable, nutrient-poor land that does not support conventional agriculture.

- Algae farms for producing biofuel can thrive without petroleum-based fertilizers, fresh water for irrigation or arable land.

- Algae can be grown rapidly at large scale and generate up to 50 times more oil per acre than other row crops like soybeans and corn.

- Algae biofuels help reduce the country's energy dependence.

- Algae use photosynthesis to capture sunlight energy for producing carbohydrates and oxygen thereby creating a natural biomass oil product.

- Algae can grow in seawater as well as high-saline water. Several species of algae can also grow in wastewater from treatment plants and water-containing phosphates, nitrates, and other contaminants.

- Algae fuels are biodegradable and non-toxic as they do not contain sulfur.

- Unlike fossil fuels, harvested algae release CO_2 when burnt, but it is absorbed by new growing algae.

Biology and Adaptation

Microalgae grow quickly and contain high oil content compared with terrestrial crops, which take a season to grow and only contain a maximum of about 5 percent dry weight of oil,. They commonly double in size every 24 hours. During the peak growth phase, some microalgae can double every three and one-half hours. Oil content of microalgae is usually between 20 percent and 50 percent, while some strains can reach as high as 80 percent. This is why microalgae are the focus in the algae-to-biofuel arena.

Table: Oil Content of Microalgae.

Microalga	Oil content (% dry weight)
Botryococcus braunii	25-75
Chlorella sp.	28-32
Crypthecodinium cohnii	20

Cylindrotheca sp.	16-37
Nitzschia sp.	45-47
Phaeodactylum tricornutum	20-30
Schizochytrium sp.	50-77
Tetraselmis suecia	15-23

Table: Oil yields Based on Crop Type.

Crop	Oil yield (gallons/acre)
Corn	18
Soybeans	48
Canola	127
Jatropha	202
Coconut	287
Oil Palm	636
Microalgae	6283-14641

Production and Agronomic Information

Most microalgae are strictly photosynthetic — that is, they need a light and carbon dioxide as energy and carbon sources. This culture mode is usually called photoautotrophic. Some algae species, however, are capable of growing in darkness and using organic carbons such as glucose or acetate as energy and carbon sources. This culture mode is termed heterotrophic. Due to high capital and operational costs, heterotrophic algal culture is hard to justify for biodiesel production. In order to minimize costs, algal biofuel production usually relies on photoautotrophic culture that uses sunlight as a free source of light.

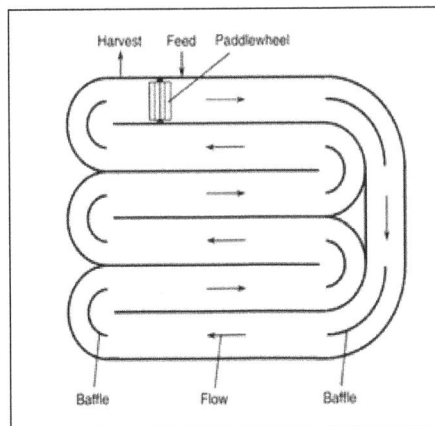

Schematic open pond system for algal culture.

Phototrophic microalgae require light, carbon dioxide, water, and inorganic salts to grow. The culture temperature should be between 15 and 30°C (~60-80°F) for optimal growth. The growth medium must contribute the inorganic elements that help make up the algal cell, such as nitrogen, phosphorus, iron, and sometimes silicon. For large-scale production of microalgae, algal cells are continuously mixed to prevent the algal biomass from settling, and nutrients are provided during daylight hours when the algae are reproducing. However, up to one-quarter of algal biomass produced during the day can be lost through respiration during the night.

A variety of photoautotrophic-based microalgal culture systems are available. For example, the algae can be grown in suspension or attached on solid surface. Each system has its own advantages and disadvantages. Currently, the suspension-based open ponds and enclosed photobioreactors are commonly used for algal biofuel production. In general, an open pond is simply a series of raceways outside, while a photobioreactor is a sophisticated reactor design which can be placed indoors in a greenhouse, or outdoors. The details of the two systems are described below.

Open ponds: Open ponds are the oldest and simplest systems for mass cultivation of microalgae. In this system, the shallow pond is usually about 1 foot deep; algae are cultured under conditions identical to their natural environment. The pond is designed in a raceway configuration, in which a paddlewheel provides circulation and mixing of the algal cells and nutrients (Figure). The raceways are typically made from poured concrete, or they are simply dug into the earth and lined with plastic to prevent the ground from soaking up the liquid. Baffles in the channel guide the flow around bends in order to minimize space. The system is often operated in a continuous mode— that is, the fresh feed containing nutrients including nitrogen phosphorus and inorganic salts is added in front of the paddle wheel. Algal broth is harvested behind the paddle wheel after it has circulated through the loop. Depending on the nutrients required by algal species, a variety of wastewater sources can be used for the algal culture, such as dairy/swine lagoon effluent and municipal wastewater. For some marine types of microalgae, seawater or water with high salinity can be used.

Although open ponds cost less to build and operate than enclosed photobioreactors, this culture system has its intrinsic disadvantages. Since these are open-air systems, they often experience a lot of water loss due to evaporation. Thus, microalgae growing in an open pond do not uptake carbon dioxide efficiently, and algal biomass production is limited. Biomass productivity is also limited by contamination with unwanted algal species as well as other organisms from feed. In addition, optimal culture conditions are difficult to maintain in open ponds, and recovering the biomass from such a dilute culture is expensive.

Enclosed photobioreactors: Enclosed photobioreactors have been employed to overcome the contamination and evaporation problems encountered in open ponds. These systems are made of transparent materials and generally placed outdoors for illumination by natural light. The cultivation vessels have a large surface area-to-volume ratio.

An algae photobioreactor.

The most widely used photobioreactor is a tubular design, which has a number of clear transparent tubes, usually aligned with the sun rays. The tubes are generally less than 10 centimeters in diameter to maximize sunlight penetration. The medium broth is circulated through a pump to the tubes, where it is exposed to light for photosynthesis, and then back to a reservoir. The algal biomass is prevented from settling by maintaining a highly turbulent flow within the reactor, using either a mechanical pump or an airlift pump. A portion of the algae is usually harvested after the solar collection tubes. In this way, continuous algal culture is possible. In some photobioreactors, the tubes are coiled spirals to form what is known as a helical tubular photobioreactor, but these sometimes require artificial illumination, which adds to the production cost. Therefore, this technology is only used for high-value products, not biodiesel feedstock.

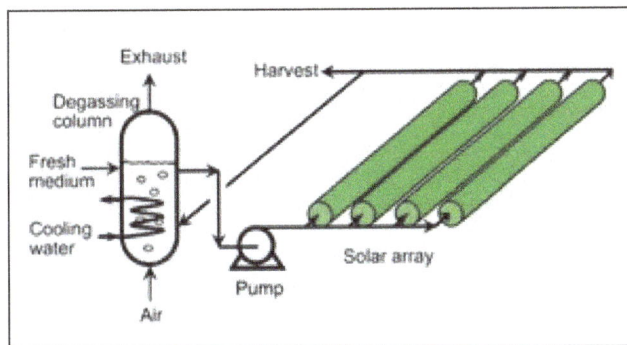

Schematic tubular photobioreactor.

The photosynthesis process generates oxygen. In an open-raceway system, this is not a problem as the oxygen is simply returned to the atmosphere. However, in the closed photobioreactor, the oxygen levels will build up until they inhibit and poison the algae. The culture must periodically be returned to a degassing zone, an area where the algal broth is bubbled with air to remove the excess oxygen. Also, the algae use carbon dioxide, which can cause carbon starvation and an increase in pH. Therefore, carbon dioxide must be fed into the system in order to successfully cultivate the microalgae on a large scale. Photobioreactors may require cooling during daylight hours, and the temperature must be regulated at night hours as well. This may be done through heat exchangers, located either in the tubes themselves or in the degassing column.

The advantages of the enclosed photobioreactors are obvious. They can overcome the problems of contamination and evaporation encountered in open ponds. The biomass productivity of photobioreactors can be 13 times greater than that of a traditional raceway pond, on average. Harvesting of biomass from photobioreactors is less expensive than that from a raceway pond, since the typical algal biomass is about 30 times as concentrated as the biomass found in raceways. However, enclosed photobioreactors also have some disadvantages. For example, the reactors are more expensive and difficult to scale up. Moreover, light limitation cannot be entirely overcome since light penetration is inversely proportional to the cell concentration. Attachment of cells to the tube walls may also prevent light penetration. Although enclosed systems can enhance the biomass concentration, the growth of microalgae is still suboptimal due to variations in temperature and light intensity.

Harvesting: After growing in open ponds or photobioreactors, the microalgae biomass needs to be harvested for further processing. The commonly used harvest method is through gravity settlement, or centrifuge. The oil from the biomass will be removed through solvent extraction and further processed into biodiesel.

Potential Yields

Depending on the culture systems used (opens ponds vs.enclosed photobioreactors), microalgae production yield is expressed as the amount of biomass per unit of surface area (for open ponds), or per unit of reactor volume (for enclosed photobioreactors). A typical open pond can produce 5 to 10 grams of biomass (dry basis) per m2 of surface area per day, which translates to 7.4 to 14.8 tons (dry biomass) per acre per year. Some researchers reported that biomass yield can be as high as 50 g/m2 per day, i.e., 74 ton biomass/m2 per year in an open pond. For enclosed photobioreactors, the biomass yield can be approximately 2 to 3 gram/L per day, i.e., 0.73-1.05 ton (dry biomass)/m3 per year. The oil content of the dry biomass is a highly variable parameter, while some strains can reach as high as 80 percent.

Production Challenges

The U.S. Department of Energy (DOE) has performed a significant effort to pursue the commercial production of algal biofuel through its ASP program from the 1980s to 1990s. After 16 years of research, DOE concluded that the algal biofuel production was still too expensive to be commercialized in the near future. Three major factors limiting commercial algal production exist: the difficulty of maintaining desirable species in the culture system, the low yield of algal oil, and the high cost of harvesting the algal biomass. DOE concluded that there was a significant amount of land, water, and CO_2 to support the algal biofuel technology.

In recent years, algal biofuel production has gained renewed interest. Both university research groups and start-up businesses are researching and developing new methods to improve the algal process efficiency with a final goal of commercial algal biofuel production. The research and development efforts can be categorized into several areas:

1. Increasing oil content of existing strains or selecting new strains with high oil content.

2. Increasing growth rate of algae.

3. Developing robust algal-growing systems in either an open-air environment or an enclosed environment.

4. Co-product development other than the oil.

5. Using algae in bioremediation.

6. Developing an efficient oil-extraction method.

One way to achieve these goals is to genetically and metabolically alter algal species. The other is to develop new or improve existing growth technologies so that the same goals listed above are met. However, it should be noted that this new wave of interest has yet to result in a significant breakthrough.

Estimated Production Cost

The production cost of the algal oil depends on many factors such as the yield of biomass from the culture system, the oil content, the scale of production systems, and the cost of recovering oil from algal biomass. Currently, algal oil production is still far more expensive than petroleum diesel fuels. For example, estimated the production cost of algae oil from a photobioreactor with an annual production capacity of 10,000 tons per year. Assuming the oil content of the algae to be around 30 percent, the author determined a production cost of $2.80/L ($10.50/gallon) of algal oil. This estimation did not include the costs of converting algal oil to biodiesel, or the distribution and marketing cost for biodiesel and taxes. At the same time, the petroleum diesel price was $2.00 to $3.00 per gallon.

Whether algal oil can be an economic source for biofuel in the future is still highly dependent on the petroleum oil price. used the following equation to estimate the cost of algal oil where it can be a competitive substitute for petroleum diesel where Calgal oil is the price of microalgal oil in $/gallon, and Cpetroleum is the price of crude oil in $/barrel:

Calgal oil = 25.9 x 10^{-3} $C_{petroleum}$

This equation assumes that algal oil has roughly 80 percent of the caloric energy value of crude petroleum. For example, with petroleum price at $100/barrel, algal oil should cost no more than $2.59/gallon in order to be competitive with petroleum diesel.

Environmental and Sustainability Issues

In addition to producing biofuel, algae can also be explored for a variety of other uses, such as fertilizer and pollution control. Certain species of algae can be land-applied for use as an organic fertilizer, either in its raw or semi-decomposed form. Algae can be grown in ponds to collect fertilizer runoff from farms; the nutrient-rich algae can then be collected and reapplied as fertilizer, potentially reducing crop-production costs. In wastewater-treatment facilities, microalgae can be used to reduce the amount of chemicals needed to clean and purify water.

In addition, algae can also be used for reducing the emissions of CO_2 from power plants. Coal is, by far, the largest fossil energy resource available in the world. About one-fourth of the world's coal reserves reside in the United States. Consumption of coal will continue to grow over the coming decades, both in the United States and the world. Through photosynthetic metabolism, microalgae absorb CO_2 and release oxygen. If an algae farm is built close to a power plant, CO_2 produced by the power plant could be utilized as a carbon source for algal growth, and the carbon emissions would be reduced by recycling waste CO_2 from power plants into clean-burning biodiesel.

Biodiesel

Biodiesel is an alternative fuel similar to conventional or 'fossil' diesel. Biodiesel can be produced from straight vegetable oil, animal oil/fats, tallow and waste cooking oil. The process used to convert these oils to Biodiesel is called transesterification. The largest possible source of suitable oil comes from oil crops such as rapeseed, palm or soybean. In the UK rapeseed represents the greatest potential for biodiesel production. Most biodiesel produced at present is produced from waste vegetable oil sourced from restaurants, chip shops, industrial food producers such as Birdseye etc. Though oil straight from the agricultural industry represents the greatest potential source it is not being produced commercially simply because the raw oil is too expensive. After the cost of converting it to biodiesel has been added on it is simply too expensive to compete with fossil diesel. Waste vegetable oil can often be sourced for free or sourced already treated for a small price. (The waste oil must be treated before conversion to biodiesel to remove impurities). The result is Biodiesel produced from waste vegetable oil can compete with fossil diesel.

Biodiesel has many environmentally beneficial properties. The main benefit of biodiesel is that it can be described as 'carbon neutral'. This means that the fuel produces no net output of carbon in the form of carbon dioxide (CO_2). This effect occurs because when the oil crop grows it absorbs the same amount of CO_2 as is released when the fuel is combusted. In fact this is not completely accurate as CO_2 is released during the production of the fertilizer required to fertilize the fields in which the oil crops are grown. Fertilizer production is not the only source of pollution associated with the production of biodiesel, other sources include the esterification process, the solvent extraction of the oil, refining, drying and transporting. All these processes require an energy input either in the form of electricity or from a fuel, both of which will generally result in the release of green house gases. To properly assess the impact of all these sources requires use of a technique called life cycle analysis. Biodiesel is rapidly biodegradable and completely non-toxic, meaning spillages represent far less of a risk than fossil diesel spillages. Biodiesel has a higher flash point than fossil diesel and so is safer in the event of a crash.

Blends

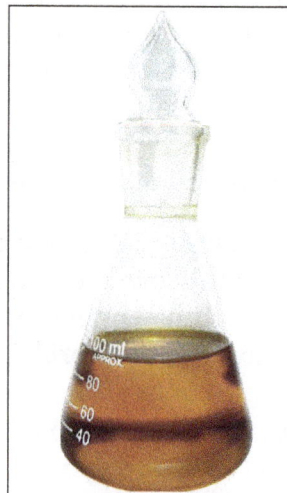

Biodiesel sample

Blends of biodiesel and conventional hydrocarbon-based diesel are products most commonly distributed for use in the retail diesel fuel marketplace. Much of the world uses a system known as the "B" factor to state the amount of biodiesel in any fuel mix:

- 100% biodiesel is referred to as B100.

- 20% biodiesel, 80% petrodiesel is labeled B20.

- 5% biodiesel, 95% petrodiesel is labeled B5.

- 2% biodiesel, 98% petrodiesel is labeled B2.

Blends of 20% biodiesel and lower can be used in diesel equipment with no, or only minor modifications, although certain manufacturers do not extend warranty coverage if equipment is damaged by these blends. The B6 to B20 blends are covered by the ASTM D7467 specification. Biodiesel can also be used in its pure form (B100), but may require certain engine modifications to avoid maintenance and performance problems. Blending B100 with petroleum diesel may be accomplished by:

- Mixing in tanks at manufacturing point prior to delivery to tanker truck.

- Splash mixing in the tanker truck (adding specific percentages of biodiesel and petroleum diesel).

- In-line mixing, two components arrive at tanker truck simultaneously.

- Metered pump mixing, petroleum diesel and biodiesel meters are set to X total volume, transfer pump pulls from two points and mix is complete on leaving pump.

Applications

Biodiesel can be used in pure form (B100) or may be blended with petroleum diesel at any concentration in most injection pump diesel engines. New extreme high-pressure (29,000 psi) common rail engines have strict factory limits of B5 or B20, depending on manufacturer. Biodiesel has different solvent properties from petrodiesel, and will degrade natural rubber gaskets and hoses in vehicles (mostly vehicles manufactured before 1992), although these tend to wear out naturally and most likely will have already been replaced with FKM, which is nonreactive to biodiesel. Biodiesel has been known to break down deposits of residue in the fuel lines where petrodiesel has been used. As a result, fuel filters may become clogged with particulates if a quick transition to pure biodiesel is made. Therefore, it is recommended to change the fuel filters on engines and heaters shortly after first switching to a biodiesel blend.

Targray Biofuels railcar transporting Biodiesel.

Distribution

Since the passage of the Energy Policy Act of 2005, biodiesel use has been increasing in the United States. In the UK, the Renewable Transport Fuel Obligation obliges suppliers to include 5% renewable fuel in all transport fuel sold in the UK by 2010. For road diesel, this effectively means 5% biodiesel (B5).

Vehicular use and Manufacturer Acceptance

In 2005, Chrysler (then part of Daimler Chrysler) released the Jeep Liberty CRD diesels from the factory into the European market with 5% biodiesel blends, indicating at least partial acceptance of biodiesel as an acceptable diesel fuel additive. In 2007, DaimlerChrysler indicated its intention to increase warranty coverage to 20% biodiesel blends if biofuel quality in the United States can be standardized.

The Volkswagen Group has released a statement indicating that several of its vehicles are compatible with B5 and B100 made from rape seed oil and compatible with the EN 14214 standard. The use of the specified biodiesel type in its cars will not void any warranty.

Mercedes Benz does not allow diesel fuels containing greater than 5% biodiesel (B5) due to concerns about "production shortcomings". Any damages caused by the use of such non-approved fuels will not be covered by the Mercedes-Benz Limited Warranty.

Starting in 2004, the city of Halifax, Nova Scotia decided to update its bus system to allow the fleet of city buses to run entirely on a fish-oil based biodiesel. This caused the city some initial mechanical issues, but after several years of refining, the entire fleet had successfully been converted.

In 2007, McDonald's of UK announced it would start producing biodiesel from the waste oil by-product of its restaurants. This fuel would be used to run its fleet.

The 2014 Chevy Cruze Clean Turbo Diesel, direct from the factory, will be rated for up to B20 (blend of 20% biodiesel/80% regular diesel) biodiesel compatibility.

Railway Usage

Biodiesel locomotive and its external fuel tank at Mount Washington Cog Railway.

British train operating company Virgin Trains claimed to have run the UK's first "biodiesel train", which was converted to run on 80% petrodiesel and 20% biodiesel.

The British Royal Train on 15 September 2007 completed its first ever journey run on 100% biodiesel fuel supplied by Green Fuels Ltd. Prince Charles and Green Fuels managing director James Hygate were the first passengers on a train fueled entirely by biodiesel fuel. Since 2007, the Royal Train has operated successfully on B100 (100% biodiesel).

Similarly, a state-owned short-line railroad in eastern Washington ran a test of a 25% biodiesel / 75% petrodiesel blend during the summer of 2008, purchasing fuel from a biodiesel producer sited along the railroad tracks. The train will be powered by biodiesel made in part from canola grown in agricultural regions through which the short line runs.

Also in 2007, Disneyland began running the park trains on B98 (98% biodiesel). The program was discontinued in 2008 due to storage issues, but in January 2009, it was announced that the park would then be running all trains on biodiesel manufactured from its own used cooking oils. This is a change from running the trains on soy-based biodiesel.

In 2007, the historic Mt. Washington Cog Railway added the first biodiesel locomotive to its all-steam locomotive fleet. The fleet has climbed up the western slopes of Mount Washington in New Hampshire since 1868 with a peak vertical climb of 37.4 degrees.

On 8 July 2014, the then Indian Railway Minister D.V. Sadananda Gowda announced in Railway Budget that 5% bio-diesel will be used in Indian Railways' Diesel Engines.

Aircraft Use

A test flight has been performed by a Czech jet aircraft completely powered on biodiesel. Other recent jet flights using biofuel, however, have been using other types of renewable fuels.

On November 7, 2011 United Airlines flew the world's first commercial aviation flight on a microbially derived biofuel using Solajet™, Solazyme's algae-derived renewable jet fuel. The Eco-skies Boeing 737-800 plane was fueled with 40 percent Solajet and 60 percent petroleum-derived jet fuel. The commercial Eco-skies flight 1403 departed from Houston's IAH airport at 10:30 and landed at Chicago's ORD airport at 13:03.

In September 2016, the Dutch flag carrier KLM contracted AltAir Fuels to supply all KLM flights departing Los Angeles International Airport with biofuel. For the next three years, the Paramount, California-based company will pump biofuel directly to the airport from their nearby refinery.

Heating Oil

Biodiesel can also be used as a heating fuel in domestic and commercial boilers, a mix of heating oil and biofuel which is standardized and taxed slightly differently from diesel fuel used for transportation. Bioheat fuel is a proprietary blend of biodiesel and traditional heating oil. Bioheat is a registered trademark of the National Biodiesel Board [NBB] and the National Oilheat Research Alliance [NORA] in the U.S., and Columbia Fuels in Canada. Heating biodiesel is available in various blends. ASTM 396 recognizes blends of up to 5 percent biodiesel as equivalent to pure petroleum heating oil. Blends of higher levels of up to 20% biofuel are used by many consumers. Research is underway to determine whether such blends affect performance.

Older furnaces may contain rubber parts that would be affected by biodiesel's solvent properties, but can otherwise burn biodiesel without any conversion required. Care must be taken, however, given that varnishes left behind by petrodiesel will be released and can clog pipes- fuel filtering and prompt filter replacement is required. Another approach is to start using biodiesel as a blend, and decreasing the petroleum proportion over time can allow the varnishes to come off more gradually and be less likely to clog. Thanks to its strong solvent properties, however, the furnace is cleaned out and generally becomes more efficient. A technical research paper describes laboratory research and field trials project using pure biodiesel and biodiesel blends as a heating fuel in oil-fired boilers. During the Biodiesel Expo 2006 in the UK, Andrew J. Robertson presented his biodiesel heating oil research from his technical paper and suggested B20 biodiesel could reduce UK household CO_2 emissions by 1.5 million tons per year.

A law passed under Massachusetts Governor Deval Patrick requires all home heating diesel in that state to be 2% biofuel by July 1, 2010, and 5% biofuel by 2013. New York City has passed a similar law.

Cleaning Oil Spills

With 80–90% of oil spill costs invested in shoreline cleanup, there is a search for more efficient and cost-effective methods to extract oil spills from the shorelines. Biodiesel has displayed its capacity to significantly dissolve crude oil, depending on the source of the fatty acids. In a laboratory setting, oiled sediments that simulated polluted shorelines were sprayed with a single coat of biodiesel and exposed to simulated tides. Biodiesel is an effective solvent to oil due to its methyl ester component, which considerably lowers the viscosity of the crude oil. Additionally, it has a higher buoyancy than crude oil, which later aids in its removal. As a result, 80% of oil was removed from cobble and fine sand, 50% in coarse sand, and 30% in gravel. Once the oil is liberated from the shoreline, the oil-biodiesel mixture is manually removed from the water surface with skimmers. Any remaining mixture is easily broken down due to the high biodegradability of biodiesel, and the increased surface area exposure of the mixture.

Biodiesel in Generators

In 2001, UC Riverside installed a 6-megawatt backup power system that is entirely fueled by biodiesel. Backup diesel-fueled generators allow companies to avoid damaging blackouts of critical operations at the expense of high pollution and emission rates. By using B100, these generators were able to essentially eliminate the byproducts that result in smog, ozone, and sulfur emissions. The use of these generators in residential areas around schools, hospitals, and the general public result in substantial reductions in poisonous carbon monoxide and particulate matter.

Biodiesel is also used in rental generators.

Properties

Biodiesel has promising lubricating properties and cetane ratings compared to low sulfur diesel fuels. Fuels with higher lubricity may increase the usable life of high-pressure fuel injection equipment that relies on the fuel for its lubrication. Depending on the engine, this might include high pressure injection pumps, pump injectors (also called *unit injectors*) and fuel injectors.

Older diesel Mercedes are popular for running on biodiesel.

The calorific value of biodiesel is about 37.27 MJ/kg. This is 9% lower than regular Number 2 petrodiesel. Variations in biodiesel energy density is more dependent on the feedstock used than the production process. Still, these variations are less than for petrodiesel. It has been claimed biodiesel gives better lubricity and more complete combustion thus increasing the engine energy output and partially compensating for the higher energy density of petrodiesel.

The color of biodiesel ranges from golden to dark brown, depending on the production method. It is slightly miscible with water, has a high boiling point and low vapor pressure. The flash point of biodiesel exceeds 130 °C (266 °F), significantly higher than that of petroleum diesel which may be as low as 52 °C (126 °F). Biodiesel has a density of ~0.88 g/cm³, higher than petrodiesel (~0.85 g/cm³).

Biodiesel contains virtually no sulfur, and it is often used as an additive to ultra-low-sulfur diesel (ULSD) fuel to aid with lubrication, as the sulfur compounds in petrodiesel provide much of the lubricity.

Fuel Efficienc

The power output of biodiesel depends on its blend, quality, and load conditions under which the fuel is burnt. The thermal efficiency for example of B100 as compared to B20 will vary due to the differing energy content of the various blends. Thermal efficiency of a fuel is based in part on fuel characteristics such as: viscosity, specific density, and flash point; these characteristics will change as the blends as well as the quality of biodiesel varies. The American Society for Testing and Materials has set standards in order to judge the quality of a given fuel sample.

One study found that the brake thermal efficiency of B40 was superior to traditional petroleum counterpart at higher compression ratios (this higher brake thermal efficiency was recorded at compression ratios of 21:1). It was noted that, as the compression ratios increased, the efficiency of all fuel types – as well as blends being tested – increased; though it was found that a blend of B40 was the most economical at a compression ratio of 21:1 over all other blends. The study implied that this increase in efficiency was due to fuel density, viscosity, and heating values of the fuels.

Combustion

Fuel systems on some modern diesel engines were not designed to accommodate biodiesel, while many heavy duty engines are able to run with biodiesel blends up to B20. Traditional direct injection fuel systems operate at roughly 3,000 psi at the injector tip while the modern common rail fuel system operates upwards of 30,000 PSI at the injector tip. Components are designed to operate at a great temperature range, from below freezing to over 1,000 °F (560 °C). Diesel fuel is expected to burn efficiently and produce as few emissions as possible. As emission standards are being introduced to diesel engines the need to control harmful emissions is being designed into the parameters of diesel engine fuel systems. The traditional inline injection system is more forgiving to poorer quality fuels as opposed to the common rail fuel system. The higher pressures and tighter tolerances of the common rail system allows for greater control over atomization and injection timing. This control of atomization as well as combustion allows for greater efficiency of modern diesel engines as well as greater control over emissions. Components within a diesel fuel system interact with the fuel in a way to ensure efficient operation of the fuel system and so the engine. If an out-of-specification fuel is introduced to a system that has specific parameters of operation, then the integrity of the overall fuel system may be compromised. Some of these parameters such as spray pattern and atomization are directly related to injection timing.

One study found that during atomization, biodiesel and its blends produced droplets greater in diameter than the droplets produced by traditional petrodiesel. The smaller droplets were attributed to the lower viscosity and surface tension of traditional diesel fuel. It was found that droplets at the periphery of the spray pattern were larger in diameter than the droplets at the center. This was attributed to the faster pressure drop at the edge of the spray pattern; there was a proportional relationship between the droplet size and the distance from the injector tip. It was found that B100 had the greatest spray penetration, this was attributed to the greater density of B100. Having a greater droplet size can lead to inefficiencies in the combustion, increased emissions, and decreased horse power. In another study it was found that there is a short injection delay when injecting biodiesel. This injection delay was attributed to the greater viscosity of Biodiesel. It was noted that the higher viscosity and the greater cetane rating of biodiesel over traditional petrodiesel lead to poor atomization, as well as mixture penetration with air during the ignition delay period. Another study noted that this ignition delay may aid in a decrease of NOx emission.

Emissions

Emissions are inherent to the combustion of diesel fuels that are regulated by the U.S. Environmental Protection Agency (E.P.A.). As these emissions are a byproduct of the combustion process, in order to ensure E.P.A. compliance a fuel system must be capable of controlling the combustion of fuels as well as the mitigation of emissions. There are a number of new technologies being phased in to control the production of diesel emissions. The exhaust gas recirculation system, E.G.R., and the diesel particulate filter, D.P.F., are both designed to mitigate the production of harmful emissions.

A study performed by the Chonbuk National University concluded that a B30 biodiesel blend reduced carbon monoxide emissions by approximately 83% and particulate matter emissions by roughly 33%. NOx emissions, however, were found to increase without the application of an E.G.R. system. The study also concluded that, with E.G.R, a B20 biodiesel blend considerably reduced the emissions of the engine. Additionally, analysis by the California Air Resources Board found that biodiesel had the lowest carbon emissions of the fuels tested, those being ultra-low-sulfur diesel, gasoline, corn-based ethanol, compressed natural gas, and five types of biodiesel from varying feedstocks. Their conclusions also showed great variance in carbon emissions of biodiesel based on the feedstock used. Of soy, tallow, canola, corn, and used cooking oil, soy showed the highest carbon emissions, while used cooking oil produced the lowest.

While studying the effect of biodiesel on diesel particulate filters, it was found that though the presence of sodium and potassium carbonates aided in the catalytic conversion of ash, as the diesel particulates are catalyzed, they may congregate inside the D.P.F. and so interfere with the clearances of the filter. This may cause the filter to clog and interfere with the regeneration process. In a study on the impact of E.G.R. rates with blends of jathropa biodiesel it was shown that there was a decrease in fuel efficiency and torque output due to the use of biodiesel on a diesel engine designed with an E.G.R. system. It was found that CO and CO_2 emissions increased with an increase in exhaust gas recirculation but NOx levels decreased. The opacity level of the jathropa blends was in an acceptable range, where traditional diesel was out of acceptable standards. It was shown that a decrease in Nox emissions could be obtained with an E.G.R. system. This study showed an advantage over traditional diesel within a certain operating range of the E.G.R. system.

As of 2017, blended biodiesel fuels (especially B5, B8, and B20) are regularly used in many heavy-duty vehicles, especially transit buses in US cities. Characterization of exhaust emissions showed significant emission reductions compared to regular diesel.

Material Compatibility

- Plastics: High-density polyethylene (HDPE) is compatible but polyvinyl chloride (PVC) is slowly degraded. Polystyrene is dissolved on contact with biodiesel.

- Metals: Biodiesel (like methanol) has an effect on copper-based materials (e.g. brass), and it also affects zinc, tin, lead, and cast iron. Stainless steels (316 and 304) and aluminum are unaffected.

- Rubber: Biodiesel also affects types of natural rubbers found in some older engine components. Studies have also found that fluorinated elastomers (FKM) cured with peroxide and base-metal oxides can be degraded when biodiesel loses its stability caused by oxidation. Commonly used synthetic rubbers FKM- GBL-S and FKM- GF-S found in modern vehicles were found to handle biodiesel in all conditions.

Technical Standards

Biodiesel has a number of standards for its quality including European standard EN 14214, ASTM International D6751, and others.

Low Temperature Gelling

When biodiesel is cooled below a certain point, some of the molecules aggregate and form crystals. The fuel starts to appear cloudy once the crystals become larger than one quarter of the wavelengths of visible light – this is the cloud point (CP). As the fuel is cooled further these crystals become larger. The lowest temperature at which fuel can pass through a 45 micrometre filter is the cold filter plugging point (CFPP). As biodiesel is cooled further it will gel and then solidify. Within Europe, there are differences in the CFPP requirements between countries. This is reflected in the different national standards of those countries. The temperature at which pure (B100) biodiesel starts to gel varies significantly and depends upon the mix of esters and therefore the feedstock oil used to produce the biodiesel. For example, biodiesel produced from low erucic acid varieties of canola seed (RME) starts to gel at approximately –10 °C (14 °F). Biodiesel produced from beef tallow and palm oil tends to gel at around 16 °C (61 °F) and 13 °C (55 °F) respectively. There are a number of commercially available additives that will significantly lower the pour point and cold filter plugging point of pure biodiesel. Winter operation is also possible by blending biodiesel with other fuel oils including #2 low sulfur diesel fuel and #1 diesel/kerosene.

Another approach to facilitate the use of biodiesel in cold conditions is by employing a second fuel tank for biodiesel in addition to the standard diesel fuel tank. The second fuel tank can be insulated and a heating coil using engine coolant is run through the tank. The fuel tanks can be switched over when the fuel is sufficiently warm. A similar method can be used to operate diesel vehicles using straight vegetable oil.

Contamination by Water

Biodiesel may contain small but problematic quantities of water. Although it is only slightly miscible with water it is hygroscopic. One of the reasons biodiesel can absorb water is the persistence of mono and diglycerides left over from an incomplete reaction. These molecules can act as an emulsifier, allowing water to mix with the biodiesel. In addition, there may be water that is residual to processing or resulting from storage tank condensation. The presence of water is a problem because:

- Water reduces the heat of fuel combustion, causing smoke, harder starting, and reduced power.

- Water causes corrosion of fuel system components (pumps, fuel lines, etc).

- Microbes in water cause the paper-element filters in the system to rot and fail, causing failure of the fuel pump due to ingestion of large particles.

- Water freezes to form ice crystals that provide sites for nucleation, accelerating gelling of the fuel.

- Water causes pitting in pistons.

Previously, the amount of water contaminating biodiesel has been difficult to measure by taking samples, since water and oil separate. However, it is now possible to measure the water content using water-in-oil sensors.

Water contamination is also a potential problem when using certain chemical catalysts involved in the production process, substantially reducing catalytic efficiency of base (high pH) catalysts such as potassium hydroxide. However, the super-critical methanol production methodology, whereby the transesterification process of oil feedstock and methanol is effectuated under high temperature and pressure, has been shown to be largely unaffected by the presence of water contamination during the production phase.

Availability and Prices

In some countries biodiesel is less expensive than conventional diesel.

Global biodiesel production reached 3.8 million tons in 2005. Approximately 85% of biodiesel production came from the European Union.

In 2007, in the United States, average retail (at the pump) prices, including federal and state fuel taxes, of B2/B5 were lower than petroleum diesel by about 12 cents, and B20 blends were the same as petro-diesel. However, as part of a dramatic shift in diesel pricing, by July 2009, the US DOE was reporting average costs of B20 15 cents per gallon higher than petroleum diesel ($2.69/gal vs. $2.54/gal). B99 and B100 generally cost more than petrodiesel except where local governments provide a tax incentive or subsidy. In the month of October 2016, Biodiesel (B20) was 2 cents lower/gallon than petrodiesel.

Production

Biodiesel is commonly produced by the transesterification of the vegetable oil or animal fat feed-stock, and other non-edible raw materials such as frying oil, etc. There are several methods for carrying out this transesterification reaction including the common batch process, heterogeneous catalysts, supercritical processes, ultrasonic methods, and even microwave methods.

Chemically, transesterified biodiesel comprises a mix of mono-alkyl esters of long chain fatty acids. The most common form uses methanol (converted to sodium methoxide) to produce methyl esters (commonly referred to as Fatty Acid Methyl Ester – FAME) as it is the cheapest alcohol available, though ethanol can be used to produce an ethyl ester (commonly referred to as Fatty Acid Ethyl Ester – FAEE) biodiesel and higher alcohols such as isopropanol and butanol have also been used. Using alcohols of higher molecular weights improves the cold flow properties of the resulting ester, at the cost of a less efficient transesterification reaction. A lipid transesterification production process is used to convert the base oil to the desired esters. Any free fatty acids (FFAs) in the base oil are either converted to soap and removed from the process, or they are esterified (yielding more biodiesel) using an acidic catalyst. After this processing, unlike straight vegetable oil, biodiesel has combustion properties very similar to those of petroleum diesel, and can replace it in most current uses.

The methanol used in most biodiesel production processes is made using fossil fuel inputs. However, there are sources of renewable methanol made using carbon dioxide or biomass as feedstock, making their production processes free of fossil fuels.

A by-product of the transesterification process is the production of glycerol. For every 1 tonne of biodiesel that is manufactured, 100 kg of glycerol are produced. Originally, there was a valuable market for the glycerol, which assisted the economics of the process as a whole. However, with the increase in global biodiesel production, the market price for this crude glycerol (containing 20% water and catalyst residues) has crashed. Research is being conducted globally to use this glycerol as a chemical building block. One initiative in the UK is The Glycerol Challenge.

Usually this crude glycerol has to be purified, typically by performing vacuum distillation. This is rather energy intensive. The refined glycerol (98%+ purity) can then be utilised directly, or converted into other products. The following announcements were made in 2007: A joint venture of Ashland Inc. and Cargill announced plans to make propylene glycol in Europe from glycerol and Dow Chemical announced similar plans for North America. Dow also plans to build a plant in China to make epichlorhydrin from glycerol. Epichlorhydrin is a raw material for epoxy resins.

Production Levels

In 2007, biodiesel production capacity was growing rapidly, with an average annual growth rate from 2002–06 of over 40%. For the year 2006, the latest for which actual production figures could be obtained, total world biodiesel production was about 5–6 million tonnes, with 4.9 million tonnes processed in Europe (of which 2.7 million tonnes was from Germany) and most of the rest from the USA. In 2008 production in Europe alone had risen to 7.8 million tonnes. In July 2009, a duty was added to American imported biodiesel in the European Union in order to balance the competition from European, especially German producers. The capacity for 2008 in Europe totalled 16 million tonnes. This compares with a total demand for diesel in the US and Europe of approximately 490 million tonnes (147 billion gallons). Total world production of vegetable oil for all purposes in 2005/06 was about 110 million tonnes, with about 34 million tonnes each of palm oil and soybean oil. As of 2018, Indonesia is the world's top supplier of palmoil-based biofuel with annual production of 3.5 million tons, and expected to export about 1 million tonnes of biodiesel.

US biodiesel production in 2011 brought the industry to a new milestone. Under the EPA Renewable Fuel Standard, targets have been implemented for the biodiesel production plants in order to monitor and document production levels in comparison to total demand. According to the year-end data released by the EPA, biodiesel production in 2011 reached more than 1 billion gallons. This production number far exceeded the 800 million gallon target set by the EPA. The projected production for 2020 is nearly 12 billion gallons.

Biodiesel Feedstocks

Plant Oils

Types

- Vegetable oil.
- Macerated oil.
- Essential oil.

Uses

- Drying oil.

- Oil paint.

- Cooking oil.

- Fuel.

- Biodiesel.

Components

- Saturated fat.

- Monounsaturated fat.

- Polyunsaturated fat.

- Trans fat.

Soybeans are used as a source of biodiesel.

A variety of oils can be used to produce biodiesel. These include:

- Virgin oil feedstock – Rapeseed and soybean oils are most commonly used, soybean oil accounting for about half of U.S. production. It also can be obtained from Pongamia, field pennycress and jatropha and other crops such as mustard, jojoba, flax, sunflower, palm oil, coconut and hemp.

- Waste vegetable oil (WVO).

- Animal fats including tallow, lard, yellow grease, chicken fat, and the by-products of the production of Omega-3 fatty acids from fish oil.

- Algae, which can be grown using waste materials such as sewage and without displacing land currently used for food production.

- Oil from halophytes such as *Salicornia bigelovii*, which can be grown using saltwater in coastal areas where conventional crops cannot be grown, with yields equal to the yields of soybeans and other oilseeds grown using freshwater irrigation.

- Sewage Sludge – The sewage-to-biofuel field is attracting interest from major companies like Waste Management and startups like InfoSpi, which are betting that renewable sewage biodiesel can become competitive with petroleum diesel on price.

Many advocates suggest that waste vegetable oil is the best source of oil to produce biodiesel, but since the available supply is drastically less than the amount of petroleum-based fuel that is burned for transportation and home heating in the world, this local solution could not scale to the current rate of consumption.

Animal fats are a by-product of meat production and cooking. Although it would not be efficient to raise animals (or catch fish) simply for their fat, use of the by-product adds value to the livestock industry (hogs, cattle, poultry). Today, multi-feedstock biodiesel facilities are producing high quality animal-fat based biodiesel. Currently, a 5-million dollar plant is being built in the US, with the intent of producing 11.4 million litres (3 million gallons) biodiesel from some of the estimated 1 billion kg (2.2 billion pounds) of chicken fat produced annually at the local Tyson poultry plant. Similarly, some small-scale biodiesel factories use waste fish oil as feedstock. An EU-funded project (ENERFISH) suggests that at a Vietnamese plant to produce biodiesel from catfish (basa, also known as pangasius), an output of 13 tons/day of biodiesel can be produced from 81 tons of fish waste (in turn resulting from 130 tons of fish). This project utilises the biodiesel to fuel a CHP unit in the fish processing plant, mainly to power the fish freezing plant.

Quantity of Feedstocks Required

Current worldwide production of vegetable oil and animal fat is not sufficient to replace liquid fossil fuel use. Furthermore, some object to the vast amount of farming and the resulting fertilization, pesticide use, and land use conversion that would be needed to produce the additional vegetable oil. The estimated transportation diesel fuel and home heating oil used in the United States is about 160 million tons (350 billion pounds) according to the Energy Information Administration, US Department of Energy. In the United States, estimated production of vegetable oil for all uses is about 11 million tons (24 billion pounds) and estimated production of animal fat is 5.3 million tonnes (12 billion pounds).

If the entire arable land area of the USA (470 million acres, or 1.9 million square kilometers) were devoted to biodiesel production from soy, this would just about provide the 160 million tonnes required (assuming an optimistic 98 US gal/acre of biodiesel). This land area could in principle be reduced significantly using algae, if the obstacles can be overcome. The US DOE estimates that if algae fuel replaced all the petroleum fuel in the United States, it would require 15,000 square miles (39,000 square kilometers), which is a few thousand square miles larger than Maryland, or 30% greater than the area of Belgium, assuming a yield of 140 tonnes/hectare (15,000 US gal/acre). Given a more realistic yield of 36 tonnes/hectare (3834 US gal/acre) the area required is about 152,000 square kilometers, or roughly equal to that of the state of Georgia or of England and Wales. The advantages of algae are that it can be grown on non-arable land such as deserts or in marine environments, and the potential oil yields are much higher than from plants.

Yield

Feedstock yield efficiency per unit area affects the feasibility of ramping up production to the huge industrial levels required to power a significant percentage of vehicles.

Some typical yields		
Crop	Yield	
	L/ha	US gal/acre
Palm oil	4752	508
Coconut	2151	230
Cyperus esculentus	1628	174
Rapeseed	954	102
Soy (Indiana)	554–922	59.2–98.6
Chinese tallow	907	97
Peanut	842	90
Sunflower	767	82
Hemp	242	26

Algae fuel yields have not yet been accurately determined, but DOE is reported as saying that algae yield 30 times more energy per acre than land crops such as soybeans. Yields of 36 tonnes/hectare are considered practical by Ami Ben-Amotz of the Institute of Oceanography in Haifa, who has been farming Algae commercially for over 20 years.

Jatropha has been cited as a high-yield source of biodiesel but yields are highly dependent on climatic and soil conditions. The estimates at the low end put the yield at about 200 US gal/acre (1.5-2 tonnes per hectare) per crop; in more favorable climates two or more crops per year have been achieved. It is grown in the Philippines, Mali and India, is drought-resistant, and can share space with other cash crops such as coffee, sugar, fruits and vegetables. It is well-suited to semi-arid lands and can contribute to slow down desertification, according to its advocates.

Efficienc and Economic Arguments

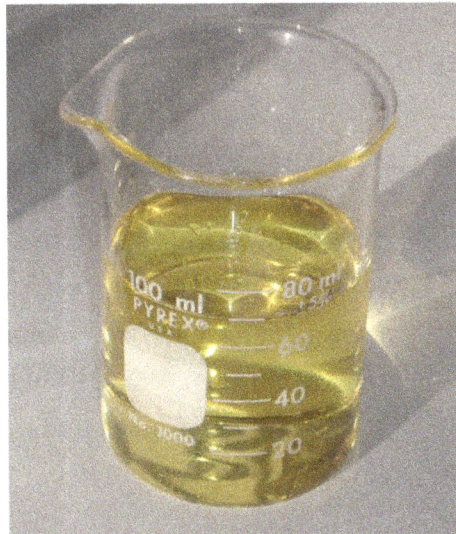

Pure biodiesel (B-100) made from soybeans.

According to a study by Drs. Van Dyne and Raymer for the Tennessee Valley Authority, the average US farm consumes fuel at the rate of 82 litres per hectare (8.75 US gal/acre) of land to produce one crop. However, average crops of rapeseed produce oil at an average rate of 1,029 L/ha (110 US gal/acre), and high-yield rapeseed fields produce about 1,356 L/ha (145 US gal/acre). The ratio of input to output in these cases is roughly 1:12.5 and 1:16.5. Photosynthesis is known to have an efficiency rate of about 3–6% of total solar radiation and if the entire mass of a crop is utilized for energy production, the overall efficiency of this chain is currently about 1% While this may compare unfavorably to solar cells combined with an electric drive train, biodiesel is less costly to deploy (solar cells cost approximately US$250 per square meter) and transport (electric vehicles require batteries which currently have a much lower energy density than liquid fuels). A 2005 study found that biodiesel production using soybeans required 27% more fossil energy than the biodiesel produced and 118% more energy using sunflowers.

However, these statistics by themselves are not enough to show whether such a change makes economic sense. Additional factors must be taken into account, such as: the fuel equivalent of the energy required for processing, the yield of fuel from raw oil, the return on cultivating food, the effect biodiesel will have on food prices and the relative cost of biodiesel versus petrodiesel, water pollution from farm run-off, soil depletion, and the externalized costs of political and military interference in oil-producing countries intended to control the price of petrodiesel.

The debate over the energy balance of biodiesel is ongoing. Transitioning fully to biofuels could require immense tracts of land if traditional food crops are used (although non food crops can be utilized). The problem would be especially severe for nations with large economies, since energy consumption scales with economic output.

If using only traditional food plants, most such nations do not have sufficient arable land to produce biofuel for the nation's vehicles. Nations with smaller economies (hence less energy consumption) and more arable land may be in better situations, although many regions cannot afford to divert land away from food production.

For third world countries, biodiesel sources that use marginal land could make more sense; e.g., pongam oiltree nuts grown along roads or jatropha grown along rail lines.

In tropical regions, such as Malaysia and Indonesia, plants that produce palm oil are being planted at a rapid pace to supply growing biodiesel demand in Europe and other markets. Scientists have shown that the removal of rainforest for palm plantations is not ecologically sound since the expansion of oil palm plantations poses a threat to natural rainforest and biodiversity.

It has been estimated in Germany that palm oil biodiesel has less than one third of the production costs of rapeseed biodiesel. The direct source of the energy content of biodiesel is solar energy captured by plants during photosynthesis. Regarding the positive energy balance of biodiesel:

When straw was left in the field, biodiesel production was strongly energy positive, yielding 1 GJ biodiesel for every 0.561 GJ of energy input (a yield/cost ratio of 1.78).

When straw was burned as fuel and oilseed rapemeal was used as a fertilizer, the yield/cost ratio for biodiesel production was even better (3.71). In other words, for every unit of energy input to produce biodiesel, the output was 3.71 units (the difference of 2.71 units would be from solar energy).

Economic Impact

Multiple economic studies have been performed regarding the economic impact of biodiesel production. One study, commissioned by the National Biodiesel Board, reported the production of biodiesel supported more than 64,000 jobs. The growth in biodiesel also helps significantly increase GDP. In 2011, biodiesel created more than $3 billion in GDP. Judging by the continued growth in the Renewable Fuel Standard and the extension of the biodiesel tax incentive, the number of jobs can increase to 50,725, $2.7 billion in income, and reaching $5 billion in GDP by 2012 and 2013.

Energy Security

One of the main drivers for adoption of biodiesel is energy security. This means that a nation's dependence on oil is reduced, and substituted with use of locally available sources, such as coal, gas, or renewable sources. Thus a country can benefit from adoption of biofuels, without a reduction in greenhouse gas emissions. While the total energy balance is debated, it is clear that the dependence on oil is reduced. One example is the energy used to manufacture fertilizers, which could come from a variety of sources other than petroleum. The US National Renewable Energy Laboratory (NREL) states that energy security is the number one driving force behind the US biofuels programme, and a White House "Energy Security for the 21st Century" paper makes it clear that energy security is a major reason for promoting biodiesel. The former EU commission president, Jose Manuel Barroso, speaking at a recent EU biofuels conference, stressed that properly managed biofuels have the potential to reinforce the EU's security of supply through diversification of energy sources.

Global Biofuel Policies

Many countries around the world are involved in the growing use and production of biofuels, such as biodiesel, as an alternative energy source to fossil fuels and oil. To foster the biofuel industry, governments have implemented legislations and laws as incentives to reduce oil dependency and to increase the use of renewable energies. Many countries have their own independent policies regarding the taxation and rebate of biodiesel use, import, and production.

Canada

It was required by the Canadian Environmental Protection Act Bill C-33 that by the year 2010, gasoline contained 5% renewable content and that by 2013, diesel and heating oil contained 2% renewable content. The EcoENERGY for Biofuels Program subsidized the production of biodiesel, among other biofuels, via an incentive rate of CAN$0.20 per liter from 2008 to 2010. A decrease of $0.04 will be applied every year following, until the incentive rate reaches $0.06 in 2016. Individual provinces also have specific legislative measures in regards to biofuel use and production.

United States

The Volumetric Ethanol Excise Tax Credit (VEETC) was the main source of financial support for biofuels, but was scheduled to expire in 2010. Through this act, biodiesel production guaranteed a tax credit of US$1 per gallon produced from virgin oils, and $0.50 per gallon made from recycled oils. Currently soybean oil is being used to produce soybean biodiesel for many commercial purposes such as blending fuel for transportation sectors.

European Union

The European Union is the greatest producer of biodiesel, with France and Germany being the top producers. To increase the use of biodiesel, there are policies requiring the blending of biodiesel into fuels, including penalties if those rates are not reached. In France, the goal was to reach 10% integration but plans for that stopped in 2010. As an incentive for the European Union countries to continue the production of the biofuel, there are tax rebates for specific quotas of biofuel produced. In Germany, the minimum percentage of biodiesel in transport diesel is set at 7% so called "B7".

Environmental Effects

The surge of interest in biodiesels has highlighted a number of environmental effects associated with its use. These potentially include reductions in greenhouse gas emissions, deforestation, pollution and the rate of biodegradation.

According to the EPA's Renewable Fuel Standards Program Regulatory Impact Analysis, released in February 2010, biodiesel from soy oil results, on average, in a 57% reduction in greenhouse gases compared to petroleum diesel, and biodiesel produced from waste grease results in an 86% reduction.

However, environmental organizations, for example, Rainforest Rescue and Greenpeace, criticize the cultivation of plants used for biodiesel production, e.g., oil palms, soybeans and sugar cane. They say the deforestation of rainforests exacerbates climate change and that sensitive ecosystems are destroyed to clear land for oil palm, soybean and sugar cane plantations. Moreover, that biofuels contribute to world hunger, seeing as arable land is no longer used for growing foods. The Environmental Protection Agency (EPA) published data in January 2012, showing that biofuels made from palm oil will not count towards the nation's renewable fuels mandate as they are not climate-friendly. Environmentalists welcome the conclusion because the growth of oil palm plantations has driven tropical deforestation, for example, in Indonesia and Malaysia.

Food, Land and Water vs. Fuel

In some poor countries the rising price of vegetable oil is causing problems. Some propose that fuel only be made from non-edible vegetable oils such as camelina, jatropha or seashore mallow which can thrive on marginal agricultural land where many trees and crops will not grow, or would produce only low yields.

Others argue that the problem is more fundamental. Farmers may switch from producing food crops to producing biofuel crops to make more money, even if the new crops are not edible. The law of supply and demand predicts that if fewer farmers are producing food the price of food will rise. It may take some time, as farmers can take some time to change which things they are growing, but increasing demand for first generation biofuels is likely to result in price increases for many kinds of food. Some have pointed out that there are poor farmers and poor countries who are making more money because of the higher price of vegetable oil.

Biodiesel from sea algae would not necessarily displace terrestrial land currently used for food production and new algaculture jobs could be created.

By comparison it should be mentioned that the production of biogas utilizes agricultural waste to generate a biofuel known as biogas, and also produces compost, thereby enhancing agriculture, sustainability and food production.

Current Research

There is ongoing research into finding more suitable crops and improving oil yield. Other sources are possible including human fecal matter, with Ghana building its first "fecal sludge-fed biodiesel plant." Using the current yields, vast amounts of land and fresh water would be needed to produce enough oil to completely replace fossil fuel usage. It would require twice the land area of the US to be devoted to soybean production, or two-thirds to be devoted to rapeseed production, to meet current US heating and transportation needs.

Specially bred mustard varieties can produce reasonably high oil yields and are very useful in crop rotation with cereals, and have the added benefit that the meal leftover after the oil has been pressed out can act as an effective and biodegradable pesticide.

The NFESC, with Santa Barbara-based Biodiesel Industries is working to develop biodiesel technologies for the US navy and military, one of the largest diesel fuel users in the world.

A group of Spanish developers working for a company called Ecofasa announced a new biofuel made from trash. The fuel is created from general urban waste which is treated by bacteria to produce fatty acids, which can be used to make biodiesel.

Another approach that does not require the use of chemical for the production involves the use of genetically modified microbes.

Algal Biodiesel

From 1978 to 1996, the U.S. NREL experimented with using algae as a biodiesel source in the "Aquatic Species Program". Michael Briggs, at the UNH Biodiesel Group, offers estimates for the realistic replacement of all vehicular fuel with biodiesel by utilizing algae that have a natural oil content greater than 50%, which Briggs suggests can be grown on algae ponds at wastewater treatment plants. This oil-rich algae can then be extracted from the system and processed into biodiesel, with the dried remainder further reprocessed to create ethanol.

The production of algae to harvest oil for biodiesel has not yet been undertaken on a commercial scale, but feasibility studies have been conducted to arrive at the above yield estimate. In addition to its projected high yield, algaculture — unlike crop-based biofuels — does not entail a decrease in food production, since it requires neither farmland nor fresh water. Many companies are pursuing algae bio-reactors for various purposes, including scaling up biodiesel production to commercial levels.

Pongamia

Millettia pinnata, also known as the Pongam Oiltree or Pongamia, is a leguminous, oilseed-bearing tree that has been identified as a candidate for non-edible vegetable oil production.

Pongamia plantations for biodiesel production have a two-fold environmental benefit. The trees both store carbon and produce fuel oil. Pongamia grows on marginal land not fit for food crops and does not require nitrate fertilizers. The oil producing tree has the highest yield of oil producing plant (approximately 40% by weight of the seed is oil) while growing in malnourished soils with high levels of salt. It is becoming a main focus in a number of biodiesel research organizations. The main advantages of Pongamia are a higher recovery and quality of oil than other crops and no direct competition with food crops. However, growth on marginal land can lead to lower oil yields which could cause competition with food crops for better soil.

Jatropha

Several groups in various sectors are conducting research on Jatropha curcas, a poisonous shrub-like tree that produces seeds considered by many to be a viable source of biodiesel feedstock oil. Much of this research focuses on improving the overall per acre oil yield of Jatropha through advancements in genetics, soil science, and horticultural practices.

SG Biofuels, a San Diego-based Jatropha developer, has used molecular breeding and biotechnology to produce elite hybrid seeds of Jatropha that show significant yield improvements over first generation varieties. SG Biofuels also claims that additional benefits have arisen from such strains, including improved flowering synchronicity, higher resistance to pests and disease, and increased cold weather tolerance.

Plant Research International, a department of the Wageningen University and Research Centre in the Netherlands, maintains an ongoing Jatropha Evaluation Project (JEP) that examines the feasibility of large scale Jatropha cultivation through field and laboratory experiments.

The Center for Sustainable Energy Farming (CfSEF) is a Los Angeles-based non-profit research organization dedicated to Jatropha research in the areas of plant science, agronomy, and horticulture. Successful exploration of these disciplines is projected to increase Jatropha farm production yields by 200–300% in the next ten years.

Fungi

A group at the Russian Academy of Sciences in Moscow published a paper in September 2008, stating that they had isolated large amounts of lipids from single-celled fungi and turned it into biodiesel in an economically efficient manner. More research on this fungal species; *Cunninghamella japonica*, and others, is likely to appear in the near future.

The recent discovery of a variant of the fungus *Gliocladium roseum* points toward the production of so-called myco-diesel from cellulose. This organism was recently discovered in the rainforests of northern Patagonia and has the unique capability of converting cellulose into medium length hydrocarbons typically found in diesel fuel.

Biodiesel From used Coffee Grounds

Researchers at the University of Nevada, Reno, have successfully produced biodiesel from oil derived from used coffee grounds. Their analysis of the used grounds showed a 10% to 15% oil content (by weight). Once the oil was extracted, it underwent conventional processing into biodiesel. It is estimated that finished biodiesel could be produced for about one US dollar per gallon. Further, it was reported that "the technique is not difficult" and that "there is so much coffee around that several hundred million gallons of biodiesel could potentially be made annually." However, even if all the coffee grounds in the world were used to make fuel, the amount produced would be less than 1 percent of the diesel used in the United States annually. "It won't solve the world's energy problem," Dr. Misra said of his work.

Exotic Sources

Recently, alligator fat was identified as a source to produce biodiesel. Every year, about 15 million pounds of alligator fat are disposed of in landfills as a waste byproduct of the alligator meat and skin industry. Studies have shown that biodiesel produced from alligator fat is similar in composition to biodiesel created from soybeans, and is cheaper to refine since it is primarily a waste product.

Biodiesel to Hydrogen-cell Power

A microreactor has been developed to convert biodiesel into hydrogen steam to power fuel cells.

Steam reforming, also known as fossil fuel reforming is a process which produces hydrogen gas from hydrocarbon fuels, most notably biodiesel due to its efficiency. A **microreactor**, or reformer, is the processing device in which water vapour reacts with the liquid fuel under high temperature and pressure. Under temperatures ranging from 700 – 1100 °C, a nickel-based catalyst enables the production of carbon monoxide and hydrogen:

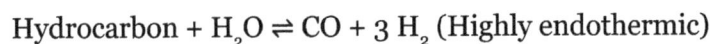

Hydrocarbon + $H_2O \rightleftharpoons CO + 3\ H_2$ (Highly endothermic)

Furthermore, a higher yield of hydrogen gas can be harnessed by further oxidizing carbon monoxide to produce more hydrogen and carbon dioxide:

$CO + H_2O \rightarrow CO_2 + H_2$ (Mildly exothermic)

Hydrogen Fuel Cells Background Information

Fuel cells operate similar to a battery in that electricity is harnessed from chemical reactions. The difference in fuel cells when compared to batteries is their ability to be powered by the constant flow of hydrogen found in the atmosphere. Furthermore, they produce only water as a by-product, and are virtually silent. The downside of hydrogen powered fuel cells is the high cost and dangers of storing highly combustible hydrogen under pressure.

One way new processors can overcome the dangers of transporting hydrogen is to produce it as necessary. The microreactors can be joined to create a system that heats the hydrocarbon under high pressure to generate hydrogen gas and carbon dioxide, a process called steam reforming. This produces up to 160 gallons of hydrogen/minute and gives the potential of powering hydrogen refueling stations, or even an on-board hydrogen fuel source for hydrogen cell vehicles. Implementation into cars would allow energy-rich fuels, such as biodiesel, to be transferred to kinetic energy while avoiding combustion and pollutant byproducts. The hand-sized square piece of metal contains microscopic channels with catalytic sites, which continuously convert biodiesel, and even its glycerol byproduct, to hydrogen.

Concerns

Engine Wear

Lubricity of fuel plays an important role in wear that occurs in an engine. A diesel engine relies on its fuel to provide lubricity for the metal components that are constantly in contact with each other. Biodiesel is a much better lubricant compared with petroleum diesel due to the presence of esters. Tests have shown that the addition of a small amount of biodiesel to diesel can significantly increase the lubricity of the fuel in short term. However, over a longer period of time (2–4 years), studies show that biodiesel loses its lubricity. This could be because of enhanced corrosion over time due to oxidation of the unsaturated molecules or increased water content in biodiesel from moisture absorption.

Fuel Viscosity

One of the main concerns regarding biodiesel is its viscosity. The viscosity of diesel is 2.5–3.2 cSt at 40 °C and the viscosity of biodiesel made from soybean oil is between 4.2 and 4.6 cSt The viscosity of diesel must be high enough to provide sufficient lubrication for the engine parts but low enough to flow at operational temperature. High viscosity can plug the fuel filter and injection system in engines. Vegetable oil is composed of lipids with long chains of hydrocarbons, to reduce its viscosity the lipids are broken down into smaller molecules of esters. This is done by converting vegetable oil and animal fats into alkyl esters using transesterification to reduce their viscosity Nevertheless, biodiesel viscosity remains higher than that of diesel, and the engine may not be able to use the fuel at low temperatures due to the slow flow through the fuel filter.

Engine Performance

Biodiesel has higher brake-specific fuel consumption compared to diesel, which means more biodiesel fuel consumption is required for the same torque. However, B20 biodiesel blend has been found to provide maximum increase in thermal efficiency, lowest brake-specific energy consumption, and lower harmful emissions. The engine performance depends on the properties of the fuel, as well as on combustion, injector pressure and many other factors. Since there are various blends of biodiesel, that may account for the contradicting reports as regards engine performance.

Vegetable Oil Fuel

Vegetable oil can be used as an alternative fuel in diesel engines and in heating oil burners. When vegetable oil is used directly as a fuel, in either modified or unmodified equipment, it is referred to as straight vegetable oil (SVO) or pure plant oil (PPO). Conventional diesel engines can be modified to help ensure that the viscosity of the vegetable oil is low enough to allow proper atomization of the fuel. This prevents incomplete combustion, which would damage the engine by causing a build-up of carbon. Straight vegetable oil can also be blended with conventional diesel or processed into biodiesel or bioliquids for use under a wider range of conditions.

Application and Usability

Modifie Fuel Systems

Most diesel car engines are suitable for the use of straight vegetable oil (SVO), also commonly called pure plant oil (PPO), with certain modifications. Principally, the viscosity and surface tension of the SVO/PPO must be reduced by preheating it, typically by using waste heat from the engine or electricity, otherwise poor atomization, incomplete combustion and carbonization may result. One common solution is to add a heat exchanger and an additional fuel tank for the petrodiesel or biodiesel blend and to switch between this additional tank and the main tank of SVO/PPO. The engine is started on diesel, switched over to vegetable oil as soon as it is warmed up and switched back to diesel shortly before being switched off to ensure that no vegetable oil remains in the engine or fuel lines when it is started from cold again. In colder climates it is often necessary to heat the vegetable oil fuel lines and tank as it can become very viscous and even solidify.

Single tank conversions have been developed, largely in Germany, which have been used throughout Europe. These conversions are designed to provide reliable operation with rapeseed oil that meets the German rapeseed oil fuel standard DIN 51605. Modifications to the engines cold start regime assist combustion on start up and during the engine warm up phase. Suitably modified indirect injection (IDI) engines have proven to be operable with 100% PPO down to temperatures of −10 °C (14 °F). Direct injection (DI) engines generally have to be preheated with a block heater or diesel fired heater. The exception is the VW Tdi (Turbocharged Direct Injection) engine for which a number of German companies offer single tank conversions. For long term durability it has been found necessary to increase the oil change frequency and to pay increased attention to engine maintenance.

Unmodifie Indirect Injection Engines

Many cars powered by indirect injection engines supplied by in-line injection pumps, or mechanical Bosch injection pumps are capable of running on pure SVO/PPO in all but winter temperatures. Indirect injection Mercedes-Benz vehicles with in-line injection pumps and cars featuring the PSA XUD engine tend to perform reasonably, especially as the latter is normally equipped with a coolant heated fuel filter. Engine reliability would depend on the condition of the engine. Attention to maintenance of the engine, particularly of the fuel injectors, cooling system and glow plugs will help to provide longevity. Ideally the engine would be converted.

Vegetable Oil Blending

The relatively high kinematic viscosity of vegetable oils must be reduced to make them compatible with conventional compression-ignition engines and fuel systems. Cosolvent blending is a low-cost and easy-to-adapt technology that reduces viscosity by diluting the vegetable oil with a low-molecular-weight solvent. This blending, or "cutting", has been done with diesel fuel, kerosene, and gasoline, amongst others; however, opinions vary as to the efficacy of this. Noted problems include higher rates of wear and failure in fuel pumps and piston rings when using blends.

Home Heating

When liquid fuels made from biomass are used for energy purposes other than transport, they are called bioliquids or biofuels.

With often minimal modification, most residential furnaces and boilers that are designed to burn No. 2 heating oil can be made to burn either biodiesel or filtered, preheated waste vegetable oil (WVO). New standard oil burners are certified to operate on 20% biodiesel (B-20). Higher blends are possible with care, since biodiesel tends to liberate existing tarry deposits in fuel tank, which may tend to clog one or more filters. Conventional oil burners tend to clog and char if more than a smaller fraction of vegetable oil is mixed with conventional oil fuel. If the vegetable oil is cleaned at home by the consumer, WVO can result in considerable savings. Many restaurants will receive a minimal amount for their used cooking oil, and processing to a biofuel is fairly simple and inexpensive. Making the oil into biodiesel involves some toxic and hazardous chemical transformations. Burning filtered WVO directly is somewhat more problematic, since it is more viscous and has a higher ignition temperature; nonetheless, its burning can be accomplished with suitable preheating or burners designed to operate on it. WVO can thus be an economical heating option for those with the necessary mechanical and experimental aptitude.

Combined Heat and Power

A number of companies offer compressed ignition engine generators optimized to run on plant oils where the waste engine heat is recovered for heating.

Properties

The main form of SVO/PPO used in the UK is rapeseed oil (also known as canola oil, primarily in the United States and Canada) which has a freezing point of −10 °C (14 °F). However the use of sunflower oil, which gels at around −12 °C (10 °F), is currently being investigated as a means of improving cold weather starting. However, oils with lower gelling points tend to be less saturated (leading to a higher iodine number) and polymerize more easily in the presence of atmospheric oxygen.

Material Compatibility

Polymerization also has been consequentially linked to catastrophic component failures such as injection pump shaft seizure and breakage, injector tip failure leading to various and/or combustion chamber components damaged. Most metallurgical problems such as corrosion and electrolysis are related to water based contamination or poor choices of plumbing (such as copper or Zinc) which can cause gelling- even with petroleum based fuels.

Temperature Effects

Some Pacific island nations are using coconut oil as fuel to reduce their expenses and their dependence on imported fuels while helping stabilize the coconut oil market. Coconut oil is only usable where temperatures do not drop below 17 degrees Celsius (63 degrees Fahrenheit), unless two-tank SVO/PPO kits or other tank-heating accessories, etc. are used. The same techniques developed to use, for example, canola and other oils in cold climates can be implemented to make coconut oil usable in temperatures lower than 17 degrees Celsius (63 degrees Fahrenheit)

Availability

Biodiesel bus.

Recycled Vegetable Oil

Recycled vegetable oil, also termed used vegetable oil (UVO), waste vegetable oil (WVO), used cooking oil (UCO), or yellow grease (in commodities exchange), is recovered from businesses and industry that use the oil for cooking.

As of 2000, the United States was producing in excess of 11 billion liters (2.9 billion U.S. gallons) of recycled vegetable oil annually, mainly from industrial deep fryers in potato processing plants, snack food factories and fast food restaurants. If all those 11 billion liters could be recycled and used to replace the energy equivalent amount of petroleum (an ideal case), almost 1% of US oil consumption could be offset. Utilizing recycled vegetable oil as a replacement for standard petroleum-derived fuels like gasoline would reduce the price of gasoline by preserving the supply of petroleum.

Virgin Vegetable Oil

Virgin vegetable oil, also termed pure plant oil or straight vegetable oil, is extracted from plants solely for use as fuel. In contrast to used vegetable oil, is not a byproduct of other industries, and thus its prospects for use as fuel are not limited by the capacities of other industries. Production of vegetable oils for use as fuels is theoretically limited only by the agricultural capacity of a given economy. However, doing so detracts from the supply of other uses of pure vegetable oil.

Legal Implications

Taxation of Fuel

Taxation on SVO/PPO as a road fuel varies from country to country. It is possible that the revenue departments in many countries are even unaware of its use or consider it too insignificant to legislate. Germany used to have 0% taxation, resulting in it being a leader in most developments of the fuel use.

Aviation Biofuel

Fuels like methanol and ethanol are not practical for aviation because they have very low energy densities. Planes would either be severely limited in their range or would not be able to take off thanks to the weight of the fuel they would need to carry. Aviation fuel has an energy density of 42 to 50 MG/kg, which is roughly the same as gasoline.

Standard Aviation Fuel

To understand what an aviation biofuel needs to be, it is important to understand what makes current aviation fuels practical. This chart lists the major properties that are required of a fuel that will be used in planes and helicopters.

Important Properties of Aviation Fuels
High Quality
Does Not Freeze
Low Risk of Explosion
High Octane
Few Contaminants

The engines that are found in aircraft come in two types: turbines and piston engines. Each requires a different kind of fuel and so the various aviation fuels will be discussed here briefly. The production of both of these fuels focuses on providing high power outputs and stable performance under the demands of flight. Of critical importance is water. Water in aviation fuel can freeze and cause lines to clog at higher altitudes. This is one of the reasons that alcohols, which tend to attract water, are not useful as aviation fuels. Cold weather performance is the most important factor in aviation fuel besides energy density.

Avgas

This is short for aviation gasoline and it is used in standard piston-engine aircraft (propellers). Its major difference from gasoline is that it is less prone to explosion, has a higher octane, and won't freeze at the low temperatures of higher altitudes. Avgas is generally being replaced by other aviation fuels. Avgas has an energy density of about 45 MG/kg.

Jet Fuel

Jet fuel is similar in many ways to kerosene and comes in two basic types: Jet A and Jet B (there are others, but we don't consider them here). Jet A fuel is designed for use in turbine engines. It is designed to rest explosion by having an autoignition temperature of 210 degrees Celsius. Jet A is subdivided into Jet A and Jet A-1. The major difference between these fuels is freezing point. Jet A fuel freezes at -40 Celsius while Jet A-1 freezes at -47 Celsius. Jet A-1 is used much more commonly than Jet A. The energy densities of these two fuels are nearly identical at about 43 MJ/kg.

Jet B fuel is "lighter" than Jet A, which means it is more prone to evaporation and thus more prone to explosion. Its major use is for cold climates where its freezing point of -60 Celsius gives it a major advantage. Jet B has an energy density of about 43 MJ/kg.

Uses of Aviation Biofuel

Despite the challenges, aviation biofuels have seen some use starting in 2008. The first flight, which was undertaken by Virgin Atlantic, used a blend of 20% biofuels. This was followed by 50-50 blends through 2012. Then in Octeober 2012, 100% biofuels was used by the National Research Council in Canada to power a Dassault Falcon 20.

Production of Aviation Biofuel

In all cases above the aviation biofuels were no different, chemically, from standard fossil fuels. It is the case that direct alcohols cannot be used as aviation fuel because they freeze easily and have low energy densities. However, alcohols can be converted to kerosene, which is the basis for all aviation fuels.

Production of kerosene from biomass can occur in several different ways. Research into the use of biological organisms is ongoing and not yet viable. Current conversion processes take the form chemical cracking and gasification, which are energy intensive and do not represent viable solutions to the large-scale production of biofuels. At this point, aviation biofuel is more of a research curiosity than a practical consideration.

Biomass Feedstock

Where the biomass for producing aviation fuel comes from plays large part in how environmentally friendly these fuels are. Both the type of plant used and the location in which it is grown are important.

Several studies have shown that using arid or former agricultural land to produce biofuels feedstock can reduce greenhouse gas emissions. Plants like Jatropha or algae can be grown in these settings and are under investigation for use as feedstock. A study from the Yale School of Forestry has shown, however, that using natural woodland to grown these plants (that is cutting down existing forest to create land for growing plants) will INCREASE greenhouse gas emissions over the use of fossil fuels.

To get an idea of just how much land would be needed to meet current demands for aviation fuel, let's consider the following graph, which shows the land areas needed if a particular feedstock is to fully replace fossil fuels in aviation and compares those to well-known land masses.

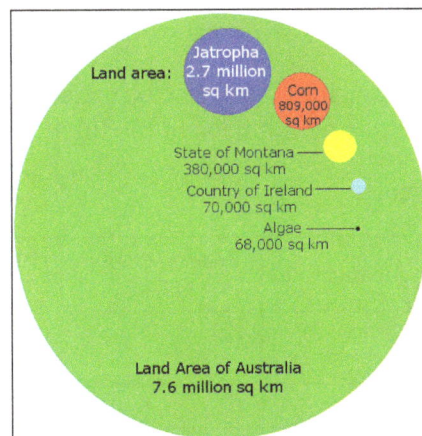

The graph above demonstrates something fantastic. Each year, 809,000 square kilometers of corn are planted, which is a land area roughly twice as large as the U.S. state of Montana. If Jatropha were used exclusively to create aviation fuel, 2.7 million square kilometers would be needed to meet current demand. Said another way, about 36% of the land area of Australia would be need to grow enough of the plant Jatropha. It is easy to see that this could have tremendous impact on land use and the food chain.

Note that algae require much less area than most other plants, which is why it is of interest to researchers. It is worth nothing, however, that little is known about what kind of impact this would have on local and global ecosystems.

Biogasoline

Biogasoline is gasoline produced from biomass such as algae. Like traditionally produced gasoline, it contains between 6 (hexane) and 12 (dodecane) carbon atoms per molecule and can be used in internal-combustion engines. Biogasoline is chemically different from biobutanol and bioethanol, as these are alcohols, not hydrocarbons.

Companies such as Diversified Energy Corporation are developing approaches to take triglyceride inputs and through a process of deoxygenation and reforming (cracking, isomerizing, aromatizing, and producing cyclic molecules) producing biogasoline. This biogasoline is intended to match the chemical, kinetic, and combustion characteristics of its petroleum counterpart, but with much higher octane levels. Others are pursuing similar approaches based on hydrotreating. And lastly still others are focused on the use of woody biomass for conversion to biogasoline using enzymatic processes.

Structure and Properties

BG100, or 100% biogasoline, can immediately be used as a drop-in substitute for petroleum gasoline in any conventional gasoline engine, and can be distributed in the same fueling infrastructure, as the properties match traditional gasoline from petroleum. Dodecane requires a small percentage of octane booster to match gasoline. Ethanol fuel (E85) requires a special engine and has lower combustion energy and corresponding fuel economy.

But due to biogasoline's chemical similarities it can also be mixed with regular gasoline. You can have higher ratios of biogasoline to gasoline and not have to modify the vehicles engine unlike ethanol.

Table: Comparison to Common Fuels.

Fuel	Energy Density MJ/L	Air-fuel ratio	Specific Energy MJ/kg	Heat of Vaporization MJ/kg	RON	MON
Gasoline	34.6	14.6	46.9	0.36	91–99	81–89
Butanol fuel	29.2	11.2	36.6	0.43	96	78
Ethanol fuel	24.0	9.0	30.0	0.92	129	102
Methanol fuel	19.7	6.5	15.6	1.2	136	104

Production

Biogasoline is created by turning sugar directly into gasoline. In late March 2010, the world's first biogasoline demonstration plant was started in Madison, WI by Virent Energy Systems, Inc. Virent discovered and developed a technique called Aqueous Phase Reforming (APR) in 2001. APR includes many processes including reforming to generate hydrogen, dehydrogenation of alcohols/hydrogenation of carbonyls, deoxygenation reactions, hydrogenolysis and cyclization. The input for APR is a carbohydrate solution created from plant material, and the product is a mixture of chemicals and oxygenated hydrocarbons. From there, the materials go through further conventional chemical processing to yield the final result: a mixture of non-oxygenated hydrocarbons that they claimed was cost-effective. These hydrocarbons are the exact hydrocarbons found in petroleum fuels which is why today's cars do not need to be altered to run on biogasoline. The only difference is in origin. Petroleum based fuels are made from oil, and biogasoline is made from plants such as beets and sugarcane or cellulosic biomass which would normally be plant waste.

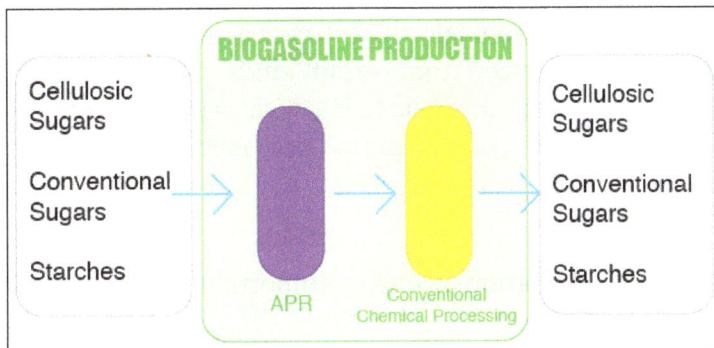

Biogasoline Production Process.

Diesel fuel is made up of linear hydrocarbons. These are long straight carbon atom chains. They differ from the shorter, branched hydrocarbons that make up gasoline. In 2014 Researchers used a feedstock of levulinic acid to create biogasoline. Levulinic acid is derived from cellulose material, such as corn stalks, straw or other plant waste. That waste does not have to be fermented. The fuel-making process is reportedly inexpensive and offers yields of over 60 percent.

Economic Viability and Future

One of the major problems facing the economic viability of biogasoline is the high up- front cost. Research groups are finding that current investment groups are impatient with the pace of biogasoline progress. In addition, environmental groups may demand that biogasoline that is produced in a way that protects wildlife, especially fish. A research group studying the economic viability of biofuels found that current techniques of production and high costs of production will prevent biogasoline from being accessible to the general public. The group determined that the price of biogasoline would need to be approximately $800 per barrel, which they determine as unlikely with current production costs. Another problem inhibiting the success of biogasoline is the lack of tax relief. The government is providing tax relief for ethanol fuels but has yet to offer tax relief for biogasoline. This makes biogasoline a much less attractive option to consumers. Lastly, producing biogasoline could have a large effect on the farming industry. If biogasoline became a serious alternative, a large percentage of our existing arable land would be converted to grow crops solely for biogasoline. This could decrease the amount of land used to farm food for human consumption and may decrease overall feedstock. This would cause an increase in overall food cost.

While there may be some problems facing the economic viability of biogasoline, the partnership between Royal Dutch Shell and Virent Energy Systems, Inc., a bioscience firm based in Madison, WI, to further research biogasoline is an encouraging sign for biogasoline's future. In addition, many nations are enacting policies that increase the use of biogasoline within the country to help curb the cost of fossil fuels and create more energy independence. Current efforts by the partnership are focused on improving the technology and making it available for large-scale production.

Biohydrogen

Biohydrogen is H_2 that is produced biologically. Interest is high in this technology because H_2 is a clean fuel and can be readily produced from certain kinds of biomass. Many challenges characterize this technology, including those intrinsic to H_2, such as storage and transportation of a non-condensible gas. Hydrogen producing organisms are poisoned by O_2. Yields of H_2 are often low.

Biochemical Principles

The main reactions involve fermentation of sugars. Important reactions start with glucose, which is converted to acetic acid:

$$C_6H_{12}O_6 + 2\ H_2O \rightarrow 2\ CH_3CO_2H + 2\ CO_2 + 4\ H_2$$

A related reaction gives formate instead of carbon dioxide:

$$C_6H_{12}O_6 + 2\ H_2O \rightarrow 2\ CH_3CO_2H + 2\ HCO_2H + 2\ H_2$$

These reactions are exergonic by 216 and 209 kcal/mol, respectively.

H_2 production is catalyzed by two hydrogenases. One is called [FeFe]-hydrogenase; the other is called [NiFe]-hydrogenase. Many organisms express these enzymes. Notable examples are members of the genera Clostridium, Desulfovibrio, Ralstonia, and the pathogen *Helicobacter*. *E. coli* is the workhorse for genetic engineering of hydrogenases.

It has been estimated that 99% of all organisms utilize dihydrogen (H_2). Most of these species are microbes and their ability to use H_2 as a metabolite arises from the expression of H_2 metalloenzymes known as hydrogenases. Hydrogenases are sub-classified into three different types based on the active site metal content: iron-iron hydrogenase, nickel-iron hydrogenase, and iron hydrogenase.

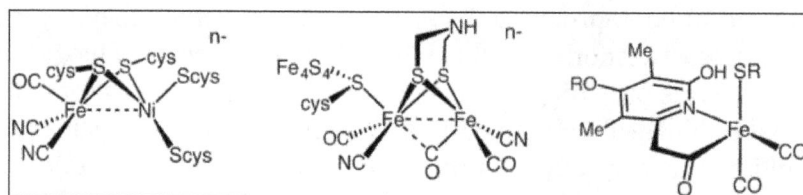

The active site structures of the three types of hydrogenase enzymes.

Production by Algae

The biological hydrogen production with algae is a method of photobiological water splitting which is done in a closed photobioreactor based on the production of hydrogen as a solar fuel by algae. Algae produce hydrogen under certain conditions. In 2000 it was discovered that if *C. reinhardtii* algae are deprived of sulfur they will switch from the production of oxygen, as in normal photosynthesis, to the production of hydrogen.

Photosynthesis

Photosynthesis in cyanobacteria and green algae splits water into hydrogen ions and electrons. The electrons are transported over ferredoxins. Fe-Fe-hydrogenases (enzymes) combine them into hydrogen gas. In *Chlamydomonas reinhardtii* Photosystem II produces in direct conversion of sunlight 80% of the electrons that end up in the hydrogen gas. Light-harvesting complex photosystem II light-harvesting protein LHCBM9 promotes efficient light energy dissipation. The Fe-Fe-hydrogenases need an anaerobic environment as they are inactivated by oxygen. Fourier transform infrared spectroscopy is used to examine metabolic pathways.

Specialized Chlorophyll

The chlorophyll (Chl) antenna size in green algae is minimized, or truncated, to maximize photobiological solar conversion efficiency and H_2 production. The truncated Chl antenna size minimizes absorption and wasteful dissipation of sunlight by individual cells, resulting in better light utilization efficiency and greater photosynthetic productivity by the green alga mass culture.

Economics

It would take about 25,000 square kilometre algal farming to produce biohydrogen equivalent to the energy provided by gasoline in the US alone. This area represents approximately 10% of the area devoted to growing soya in the US.

Bioreactor Design Issues

- Restriction of photosynthetic hydrogen production by accumulation of a proton gradient.
- Competitive inhibition of photosynthetic hydrogen production by carbon dioxide.
- Requirement for bicarbonate binding at photosystem II (PSII) for efficient photosynthetic activity.
- Competitive drainage of electrons by oxygen in algal hydrogen production.
- Economics must reach competitive price to other sources of energy and the economics are dependent on several parameters.
- A major technical obstacle is the efficiency in converting solar energy into chemical energy stored in molecular hydrogen.

Attempts are in progress to solve these problems via bioengineering.

Industrial Hydrogen

Competing for biohydrogen, at least for commercial applications, are many mature industrial processes. Hydrogen is usually derived from fossil fuels by steam reforming of natural gas - sometimes referred to as steam methane reforming (SMR) - is the most common method of producing bulk hydrogen at about 95% of the world production.

$$CH_4 + H_2O \rightleftharpoons CO + 3\,H_2$$

Bioethers

Ethers are a class of carbon compounds that contain an ether group, which looks as follows.

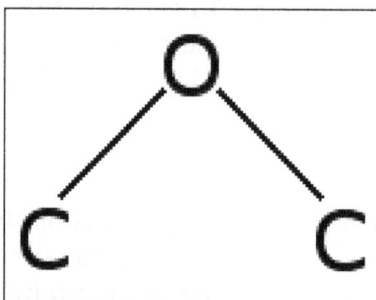

Either carbon in the basic ether above may be attached to other carbons as well. So, this image represents the most basic ether. All that is needed to make an ether is an oxygen atom with a hydrocarbon chain on either side.

Uses

Ethers are currently used as substitutes for lead to improve engine performance. They also decrease oxygen wear and toxic emissions, particularly ozone. The most commonly used fuel additive ethers are methyl-tertiary-butyl-ether (MBTE) and ethyl-tertiary-butyl-ether (ETBE). You are likely to see these abbreviations on petrol pumps.

Bioether

Bioether is made from either wheat or sugar beet. It can also be produced from the waste glycerol that results from the production of biodiesel. Bioether is likely to replace petro-ether as an additive to current fossil fuels. In the future, it is unlikely to be a fuel in and of itself because it has a low energy density. In fact, its energy density is about half that of standard diesel, which means you would need twice the volume of ether to go the same distance. Mostly likely, it will become an additive to other biofuels to reduce emissions.

	Dimethyl Ether	Diesel
Energy Density (MJ/kg)	28.8	42.7
Cetane Number	55-60	40-55

Despite its drawbacks, there is some interest in using ether, particularly dimethyl ether, as a replacement for diesel fuel. This is made possible by the fact that ether works well in compression ignition engines. The interest stems from the reduced particulate emissions and reduced pollution of ether. In urban environments, where refueling frequently is possible and pollution levels run high, the tradeoff of reduced emissions for more frequent refueling is an attractive one. The only major problem to ether in these settings is that it must be compressed to be a liquid. At standard temperatures and pressures, ether is a gas. To make it transportable, it must be compressed. Of course, compression has benefits as well. If the gas can be compressed enough, then more can be carried and refueling intervals will be longer.

Sustainable Biofuel

Sustainable biofuel is biofuel produced in a sustainable manner.

Sustainability Standards

In 2008, the Roundtable for Sustainable Biofuels released its proposed standards for sustainable biofuels. This includes 12 principles:

1. Biofuel production shall follow international treaties and national laws regarding such things as air quality, water resources, agricultural practices, labor conditions, and more.

2. Biofuels projects shall be designed and operated in participatory processes that involve all relevant stakeholders in planning and monitoring.

3. Biofuels shall significantly reduce greenhouse gas emissions as compared to fossil fuels. The principle seeks to establish a standard methodology for comparing greenhouse gases (GHG) benefits.

4. Biofuel production shall not violate human rights or labor rights, and shall ensure decent work and the well-being of workers.

5. Biofuel production shall contribute to the social and economic development of local, rural and indigenous peoples and communities.

6. Biofuel production shall not impair food security.

7. Biofuel production shall avoid negative impacts on biodiversity, ecosystems and areas of high conservation value.

8. Biofuel production shall promote practices that improve soil health and minimize degradation.

9. Surface and groundwater use will be optimized and contamination or depletion of water resources minimized.

10. Air pollution shall be minimized along the supply chain.

11. Biofuels shall be produced in the most cost-effective way, with a commitment to improve production efficiency and social and environmental performance in all stages of the biofuel value chain.

12. Biofuel production shall not violate land rights.

Several countries and regions have introduced policies or adopted standards to promote sustainable biofuels production and use, most prominently the European Union and the United States. The 2009 EU Renewable Energy Directive, which requires 10 percent of transportation energy from renewable energy by 2020, is the most comprehensive mandatory sustainability standard in place as of 2010.

The EU Renewable Energy Directive requires that the lifecycle greenhouse gas emissions of biofuels consumed be at least 50 percent less than the equivalent emissions from gasoline or diesel by 2017 (and 35 percent less starting in 2011). Also, the feedstocks for biofuels "should not be harvested from lands with high biodiversity value, from carbon-rich or forested land, or from wetlands".

As with the EU, the U.S. Renewable Fuel Standard (RFS) and the California Low Carbon Fuel Standard (LCFS) both require specific levels of lifecycle greenhouse gas reductions compared to equivalent fossil fuel consumption. The RFS requires that at least half of the biofuels production mandated by 2022 should reduce lifecycle emissions by 50 percent. The LCFS is a performance standard that calls for a minimum of 10 percent emissions reduction per unit of transport energy by 2020. Both the U.S. and California standards currently address only greenhouse gas emissions, but California plans to "expand its policy to address other sustainability issues associated with liquid biofuels in the future".

In 2009, Brazil also adopted new sustainability policies for sugarcane ethanol, including "zoning regulation of sugarcane expansion and social protocols".

Need for Sustainability

Biofuels, in the form of liquid fuels derived from plant materials, are entering the market, driven by factors such as oil price spikes and the need for increased energy security. However, many of these first-generation biofuels that are currently being supplied have been criticised for their adverse impacts on the natural environment, food security, and land use.

The challenge is to support second, third and fourth-generation biofuel development. Second-generation biofuels include new cellulosic technologies, with responsible policies and economic instruments to help ensure that biofuel commercialization is sustainable. Responsible commercialization of biofuels represents an opportunity to enhance sustainable economic prospects in Africa, Latin America and Asia.

Biofuels have a limited ability to replace fossil fuels and should not be regarded as a 'silver bullet' to deal with transport emissions. However, they offer the prospect of increased market competition and oil price moderation. A healthy supply of alternative energy sources will help to combat gasoline price spikes and reduce dependency on fossil fuels, especially in the transport sector. Using transportation fuels more efficiently is also an integral part of a sustainable transport strategy.

Biofuel Options

Biofuel development and use is a complex issue because there are many biofuel options which are available. Biofuels, such as ethanol and biodiesel, are currently produced from the products of conventional food crops such as the starch, sugar and oil feedstocks from crops that include wheat, maize, sugar cane, palm oil and oilseed rape. Some researchers fear that a major switch to biofuels from such crops would create a direct competition with their use for food and animal feed, and claim that in some parts of the world the economic consequences are already visible, other researchers look at the land available and the enormous areas of idle and abandoned land and claim that there is room for a large proportion of biofuel also from conventional crops.

Second generation biofuels are now being produced from a much broader range of feedstocks including the cellulose in dedicated energy crops (perennial grasses such as switchgrass and Miscanthus giganteus), forestry materials, the co-products from food production, and domestic vegetable waste. Advances in the conversion processes will improve the sustainability of biofuels, through better efficiencies and reduced environmental impact of producing biofuels, from both existing food crops and from cellulosic sources. One promising development in biobutanol production technology was discovered in the late summer of 2011—Tulane University's alternative fuel research scientists discovered a strain of *Clostridium* bacteria, called "TU-103", a key feature of the discovery is that the "TU-103" organism can convert nearly any form of cellulose into butanol, and is the only known strain of *Clostridium*-genus bacteria that can do so in the presence of oxygen. The university's researchers have stated that the source of the "TU-103" *Clostridium* bacteria strain was most likely from the solid waste from one of the plains zebra at New Orleans' Audubon Zoo.

In 2007, Ronald Oxburgh suggested in *The Courier-Mail* that production of biofuels could be either responsible or irresponsible and had several trade-offs: "Produced responsibly they are a sustainable energy source that need not divert any land from growing food nor damage the environment; they can also help solve the problems of the waste generated by Western society; and they can create jobs for the poor where previously were none. Produced irresponsibly, they at best offer no climate benefit and, at worst, have detrimental social and environmental consequences. In other words, biofuels are pretty much like any other product. In 2008 the Nobel prize-winning chemist Paul J. Crutzen published findings that the release of nitrous oxide (N_2O) emissions in the production of biofuels means that they contribute more to global warming than the fossil fuels they replace.

According to the Rocky Mountain Institute, sound biofuel production practices would not hamper food and fibre production, nor cause water or environmental problems, and would enhance soil fertility. The selection of land on which to grow the feedstocks is a critical component of the ability of biofuels to deliver sustainable solutions. A key consideration is the minimisation of biofuel competition for prime cropland.

Biofuels are different from fossil fuels in regard to carbon emissions being short term, but are similar to fossil fuels in that biofuels contribute to air pollution. Raw biofuels burned to generate steam for heat and power, produces airborne carbon particulates, carbon monoxide and nitrous oxides. The WHO estimates 3.7 million premature deaths worldwide in 2012 due to air pollution.

Plants used as Sustainable Biofuel

Sugarcane in Brazil

Brazil's production of ethanol fuel from sugarcane dates back to the 1970s, as a governmental response to the 1973 oil crisis. Brazil is considered the biofuel industry leader and the world's first sustainable biofuels economy.Inslee, Jay; Bracken Hendricks. "6. Homegrown Energy". Apollo's Fire. Island Press, Washington, D.C. In 2010 the U.S. Environmental Protection Agency designated Brazilian sugarcane ethanol as an advanced biofuel due to EPA's estimated 61% reduction of total life cycle greenhouse gas emissions, including direct indirect land use change emissions. Brazil sugarcane ethanol fuel program success and sustainability is based on the most efficient agricultural technology for sugarcane cultivation in the world, uses modern equipment and cheap sugar cane as feedstock, the residual cane-waste (bagasse) is used to process heat and power, which results in a very competitive price and also in a high energy balance (output energy/input energy), which varies from 8.3 for average conditions to 10.2 for best practice production.

Sugarcane (*Saccharum officinarum*) plantation ready for harvest.

Mechanized harvesting of sugarcane.

A report commissioned by the United Nations, based on a detailed review of published research up to mid-2009 as well as the input of independent experts world-wide, found that ethanol from sugar cane as produced in Brazil *"in some circumstances does better than just "zero emission". If grown and processed correctly, it has negative emission, pulling CO2 out of the atmosphere, rather than adding it.* In contrast, the report found that U.S. use of maize for biofuel is less efficient, as sugarcane can lead to emissions reductions of between 70% and well over 100% when substituted for gasoline. Several other studies have shown that sugarcane-based ethanol reduces greenhouse gases by 86 to 90% if there is no significant land use change.

In another study commissioned by the Dutch government in 2006 to evaluate the sustainability of Brazilian bioethanol concluded that there is sufficient water to supply all foreseeable long-term water requirements for sugarcane and ethanol production. This evaluation also found that consumption of agrochemicals for sugar cane production is lower than in citric, corn, coffee and soybean cropping. The study found that development of resistant sugar cane varieties is a crucial aspect of disease and pest control and is one of the primary objectives of Brazil's cane genetic improvement programs. Disease control is one of the main reasons for the replacement of a commercial variety of sugar cane.

Cosan's Costa Pinto sugar cane mill and ethanol distillery plant.

Another concern is the fact that sugarcane fields are traditionally burned just before harvest to avoid harm to the workers, by removing the sharp leaves and killing snakes and other harmful animals, and also to fertilize the fields with ash. Mechanization will reduce pollution from burning fields and has higher productivity than people, and due to mechanization the number of temporary workers in the sugarcane plantations has already declined. By the 2008 harvest season, around 47% of the cane was collected with harvesting machines.

Regarding the negative impacts of the potential direct and indirect effect of land use changes on carbon emissions, the study commissioned by the Dutch government concluded that "it is very difficult to determine the indirect effects of further land use for sugar cane production (i.e. sugar cane replacing another crop like soy or citrus crops, which in turn causes additional soy plantations replacing pastures, which in turn may cause deforestation), and also not logical to attribute all these soil carbon losses to sugar cane". The Brazilian agency Embrapa estimates that there is enough agricultural land available to increase at least 30 times the existing sugarcane plantation without endangering sensible ecosystems or taking land destined for food crops. Most future growth is expected to take place on abandoned pasture lands, as it has been the historical trend in São Paulo state. Also, productivity is expected to improve even further based on current biotechnology research, genetic improvement, and better agronomic practices, thus contributing to reduce land demand for future sugarcane cultures.

Another concern is the risk of clearing rain forests and other environmentally valuable land for sugarcane production, such as the Amazon rainforest, the Pantanal or the Cerrado. Embrapa has rebutted this concern explaining that 99.7% of sugarcane plantations are located at least 2,000 km from the Amazon, and expansion during the last 25 years took place in the Center-South region, also far away from the Amazon rainforest, the Pantanal or the Atlantic forest. In São Paulo state growth took place in abandoned pasture lands. The impact assessment commissioned by the Dutch government supported this argument.

In order to guarantee a sustainable development of ethanol production, in September 2009 the government issued by decree a countrywide agroecological land use zoning to restrict sugarcane growth in or near environmentally sensitive areas. According to the new criteria, 92.5% of the

Brazilian territory is not suitable for sugarcane plantation. The government considers that the suitable areas are more than enough to meet the future demand for ethanol and sugar in the domestic and international markets foreseen for the next decades.

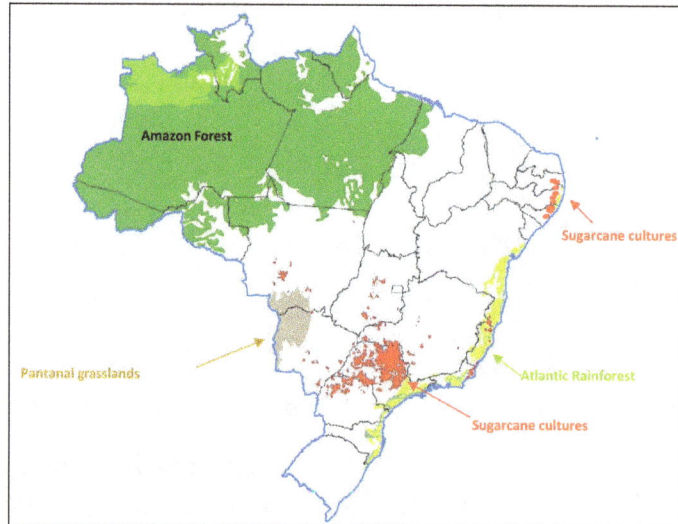

Location of environmentally valuable areas with respect to sugarcane plantations.
São Paulo, located in the Southeast Region of Brazil, concentrates two-thirds of sugarcane cultures.

Regarding the food vs fuel issue, a World Bank research report published on July 2008 found that *"Brazil's sugar-based ethanol did not push food prices appreciably higher"*. This research paper also concluded that Brazil's sugar cane–based ethanol has not raised sugar prices significantly. An economic assessment report also published in July 2008 by the OECD agrees with the World Bank report regarding the negative effects of subsidies and trade restrictions, but found that the impact of biofuels on food prices are much smaller. A study by the Brazilian research unit of the Fundação Getúlio Vargas regarding the effects of biofuels on grain prices concluded that the major driver behind the 2007-2008 rise in food prices was speculative activity on futures markets under conditions of increased demand in a market with low grain stocks. The study also concluded that there is no correlation between Brazilian sugarcane cultivated area and average grain prices, as on the contrary, the spread of sugarcane was accompanied by rapid growth of grain crops in the country.

Jatropha

India and Africa

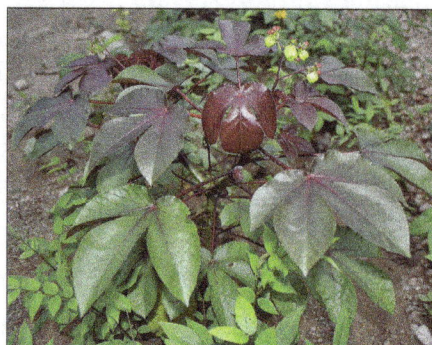

Jatropha gossipifolia.

Crops like Jatropha, used for biodiesel, can thrive on marginal agricultural land where many trees and crops won't grow, or would produce only slow growth yields. Jatropha cultivation provides benefits for local communities:

Cultivation and fruit picking by hand is labour-intensive and needs around one person per hectare. In parts of rural India and Africa this provides much-needed jobs - about 200,000 people world-wide now find employment through jatropha. Moreover, villagers often find that they can grow other crops in the shade of the trees. Their communities will avoid importing expensive diesel and there will be some for export too.

Cambodia

Cambodia has no proven fossil fuel reserves, and is almost completely dependent on imported diesel fuel for electricity production. Consequently, Cambodians face an insecure supply and pay some of the highest energy prices in the world. The impacts of this are widespread and may hinder economic development.

Biofuels may provide a substitute for diesel fuel that can be manufactured locally for a lower price, independent of the international oil price. The local production and use of biofuel also offers other benefits such as improved energy security, rural development opportunities and environmental benefits. The Jatropha curcas species appears to be a particularly suitable source of biofuel as it already grows commonly in Cambodia. Local sustainable production of biofuel in Cambodia, based on the Jatropha or other sources, offers good potential benefits for the investors, the economy, rural communities and the environment.

Mexico

Jatropha is native to Mexico and Central America and was likely transported to India and Africa in the 1500s by Portuguese sailors convinced it had medicinal uses. In 2008, recognizing the need to diversify its sources of energy and reduce emissions, Mexico passed a law to push developing biofuels that don't threaten food security and the agriculture ministry has since identified some 2.6 million hectares (6.4 million acres) of land with a high potential to produce jatropha. The Yucatán Peninsula, for instance, in addition to being a corn-producing region, also contains abandoned sisal plantations, where the growing of Jatropha for biodiesel production would not displace food.

On April 1, 2011 Interjet completed the first Mexican aviation biofuels test flight on an Airbus A320. The fuel was a 70:30 traditional jet fuel biojet blend produced from Jatropha oil provided by three Mexican producers, Global Energías Renovables (a wholly owned subsidiary of U.S.-based Global Clean Energy Holdings, Bencafser S.A. and Energy JH S.A. Honeywell's UOP processed the oil into Bio-SPK (Synthetic Paraffinic Kerosene). Global Energías Renovables operates the largest Jatropha farm in the Americas.

On August 1, 2011 Aeromexico, Boeing, and the Mexican Government participated in the first biojet powered transcontinental flight in aviation history. The flight from Mexico City to Madrid used a blend of 70 percent traditional fuel and 30 percent biofuel (aviation biofuel). The biojet was produced entirely from Jatropha oil.

Pongamia Pinnata in Australia and India

Pongamia pinnata is a legume native to Australia, India, Florida (USA) and most tropical regions, and is now being invested in as an alternative to Jatropha for areas such as Northern Australia, where Jatropha is classed as a noxious weed. Commonly known as simply 'Pongamia', this tree is currently being commercialised in Australia by Pacific Renewable Energy, for use as a Diesel replacement for running in modified Diesel engines or for conversion to Biodiesel using 1st or 2nd Generation Biodiesel techniques, for running in unmodified Diesel engines.

Pongamia pinnata seeds.

Sweet Sorghum in India

Sweet sorghum overcomes many of the shortcomings of other biofuel crops. With sweet sorghum, only the stalks are used for biofuel production, while the grain is saved for food or livestock feed. It is not in high demand in the global food market, and thus has little impact on food prices and food security. Sweet sorghum is grown on already-farmed drylands that are low in carbon storage capacity, so concerns about the clearing of rainforest do not apply. Sweet sorghum is easier and cheaper to grow than other biofuel crops in India and does not require irrigation, an important consideration in dry areas. Some of the Indian sweet sorghum varieties are now grown in Uganda for ethanol production.

A study by researchers at the International Crops Research Institute for the Semi-Arid Tropics (ICRISAT) found that growing sweet sorghum instead of grain sorghum could increase farmers incomes by US$40 per hectare per crop because it can provide food, feed and fuel. With grain sorghum currently grown on over 11 million hectares (ha) in Asia and on 23.4 million ha in Africa, a switch to sweet sorghum could have a considerable economic impact.

International Collaboration on Sustainable Biofuels

Roundtable on Sustainable Biomaterials

Public attitudes and the actions of key stakeholders can play a crucial role in realising the potential of sustainable biofuels. Informed discussion and dialogue, based both on scientific research and an understanding of public and stakeholder views, is important.

The Roundtable on Sustainable Biofuels is an international initiative which brings together farmers, companies, governments, non-governmental organizations, and scientists who are interested in the sustainability of biofuels production and distribution. During 2008, the Roundtable used meetings, teleconferences, and online discussions to develop a series of principles and criteria for sustainable biofuels production.

In April 2011, the Roundtable on Sustainable Biofuels launched a set of comprehensive sustainability criteria - the "RSB Certification System." Biofuels producers that meet to these criteria are able to show buyers and regulators that their product has been obtained without harming the environment or violating human rights.

Sustainable Biofuels Consensus

The Sustainable Biofuels Consensus is an international initiative which calls upon governments, the private sector, and other stakeholders to take decisive action to ensure the sustainable trade, production, and use of biofuels. In this way biofuels may play a key role in energy sector transformation, climate stabilization, and resulting worldwide revitalisation of rural areas.

The Sustainable Biofuels Consensus envisions a "landscape that provides food, fodder, fiber, and energy, which offers opportunities for rural development; that diversifies energy supply, restores ecosystems, protects biodiversity, and sequesters carbon".

Better Sugarcane Initiative/Bonsucro

In 2008, a multi-stakeholder process was initiated by the World Wildlife Fund and the International Finance Corporation, the private development arm of the World Bank, bringing together industry, supply chain intermediaries, end-users, farmers and civil society organisations to develop standards for certifying the derivative products of sugar cane, one of which is ethanol fuel.

The Bonsucro standard is based around a definition of sustainability which is founded on five principles:

1. Obey the law.

2. Respect human rights and labour standards.

3. Manage input, production and processing efficiencies to enhance sustainability.

4. Actively manage biodiversity and ecosystem services.

5. Continuously improve key areas of the business.

Biofuel producers that wish to sell products marked with the Bonsucro standard must both ensure that they product to the Production Standard, and that their downstream buyers meet the Chain of Custody Standard. In addition, if they wish to sell to the European market and count against the EU Renewable Energy Directive, then they must adhere to the Bonsucro EU standard, which includes specific greenhouse gas calculations following European Commission calculation guidelines.

Oil Price Moderation

Biofuels offer the prospect of real market competition and oil price moderation. According to the Wall Street Journal, crude oil would be trading 15 per cent higher and gasoline would be as much as 25 per cent more expensive, if it were not for biofuels. A healthy supply of alternative energy sources will help to combat gasoline price spikes.

Sustainable Transport

Biofuels have a limited ability to replace fossil fuels and should not be regarded as a 'silver bullet' to deal with transport emissions. Biofuels on their own cannot deliver a sustainable transport system and so must be developed as part of an integrated approach, which promotes other renewable energy options and energy efficiency, as well as reducing the overall energy demand and need for transport. Consideration needs to be given to the development of hybrid and fuel cell vehicles, public transport, and better town and rural planning.

In December 2008 an Air New Zealand jet completed the world's first commercial aviation test flight partially using jatropha-based fuel. More than a dozen performance tests were undertaken in the two-hour test flight which departed from Auckland International Airport. A biofuel blend of 50:50 jatropha and Jet A1 fuel was used to power one of the Boeing 747-400's Rolls-Royce RB211 engines. Air New Zealand set several criteria for its jatropha, requiring that "the land it came from was neither forest nor virgin grassland in the previous 20 years, that the soil and climate it came from is not suitable for the majority of food crops and that the farms are rain fed and not mechanically irrigated". The company has also set general sustainability criteria, saying that such biofuels must not compete with food resources, that they must be as good as traditional jet fuels, and that they should be cost competitive.

In January 2009, Continental Airlines used a sustainable biofuel to power a commercial aircraft for the first time in North America. This demonstration flight marks the first sustainable biofuel demonstration flight by a commercial carrier using a twin-engined aircraft, a Boeing 737-800, powered by CFM International CFM56-7B engines. The biofuel blend included components derived from algae and jatropha plants. The algae oil was provided by Sapphire Energy, and the jatropha oil by Terasol Energy.

In March 2011, Yale University research showed significant potential for sustainable aviation fuel based on jatropha-curcas. According to the research, if cultivated properly, "jatropha can deliver many benefits in Latin America and greenhouse gas reductions of up to 60 percent when compared to petroleum-based jet fuel". Actual farming conditions in Latin America were assessed using sustainability criteria developed by the Roundtable on Sustainable Biofuels. Unlike previous research, which used theoretical inputs, the Yale team conducted many interviews with jatropha farmers and used "field measurements to develop the first comprehensive sustainability analysis of actual projects".

As of June 2011, revised international aviation fuel standards officially allow commercial airlines to blend conventional jet fuel with up to 50 percent biofuels. The renewable fuels "can be blended with conventional commercial and military jet fuel through requirements in the newly issued edition of ASTM D7566, Specification for Aviation Turbine Fuel Containing Synthesized Hydrocarbons".

In December 2011, the FAA awarded $7.7 million to eight companies to advance the development of commercial aviation biofuels, with a special focus on alcohol to jet fuel. The FAA is assisting in the development of a sustainable fuel (from alcohols, sugars, biomass, and organic matter such as pyrolysis oils) that can be "dropped in" to aircraft without changing current practices and infrastructure. The research will test how the new fuels affect engine durability and quality control standards.

GreenSky London, a biofuels plant under construction in 2014, aimed to take in some 500,000 tonnes of municipal rubbish and change the organic component into 60,000 tonnes of jet fuel, and 40 megawatts of power. By the end of 2015, it was hoped all British Airways flights from London City Airport will be fuelled by waste and rubbish discarded by London residents, leading to carbon savings equivalent to taking 150,000 cars off the road. Unfortunately, the £340m scheme was mothballed in January 2016 following low crude oil prices, jittery investors and a lack of support from the UK government.

References

- How-is-biogas-produced, Biogas, biogas, About-gas: gasum.com, Retrieved , 17 February, 2019

- "Liquid Fuels - Conversion of Methanol to Gasoline". Gasifipedia. National Energy Technology Laboratory, U.S. Department of Energy. Retrieved 25 July 2014

- What-is-syngas: biofuel.org.uk, Retrieved 22 August, 2019

- Owen, K., Coley., C.S. Weaver, "Automotive Fuels Reference Book", SAE International, ISBN 978-1-56091-589-8

- Non-Methane hydrocarbons "Archived copy". Archived from the original on 2011-07-27. Retrieved 2010-11-27

- Burton, George; Holman, John; Lazonby, John (2000). Salters Advanced Chemistry: Chemical Storylines (2nd ed.). Heinemann. ISBN 0-435-63119-5

- Ethanol-fuel: conserve-energy-future.com, Retrieved 12 August, 2019

- Abraciclo (27 January 2010). "Motos flex foram as mais vendidas em 2009 na categoria 150cc" (in Portuguese). UNICA. Archived from the original on 5 December 2012. Retrieved 10 February 2010

- Ethanol-fuel: conserve-energy-future.com, Retrieved 9 May, 2019

- Dirk Lammers (2007-03-04). "Gasification may be key to U.S. Ethanol". CBS News. Archived from the original on 2007-11-22. Retrieved 2007-11-28

- Algae-for-biofuel-production: extension.org, Retrieved 1 June, 2019

- "Civic amenity site as collection point for WVO ref 2". Archived from the original on 2013-07-29. Retrieved 2013-09-17

- What-biodiesel, biofuels: strath.ac.uk, Retrieved 10 May, 2019

Feedstocks for Biofuels 4

- **Feedstocks**
- **Energy Crop**
- **Camelina**
- **Jatropha Curcas**
- **Lignocellulosic Biomass**
- **Switchgrass**
- **Miscanthus Giganteus**

Feedstock is the unprocessed material that is used to supply a manufacturing process. Some of the common feedstocks are Camelina, Jatropha Curcas, Lignocellulosic Biomass, Switchgrass and Miscanthus Giganteus. The diverse use of these energy crops as feedstocks for biofuels have been thoroughly discussed in this chapter.

Feedstocks

A feedstock refers to any unprocessed material used to supply a manufacturing process. Feedstocks are bottleneck assets because their availability determines the ability to make products.

In its most general sense, a feedstock is a natural material (e.g., ore, wood, seawater, coal) that has been transformed for marketing in large volumes.

In engineering, particularly as it relates to energy, a feedstock refers specifically to a renewable, biological material that can be converted into energy or fuel.

In chemistry, a feedstock is a chemical used to support a large-scale chemical reaction. The term usually refers to an organic substance.

A feedstock may also be called a raw material or unprocessed material. Sometimes feedstock is a synonym for biomass.

Examples of Feedstocks

Using the broad definition of a feedstock, any natural resource might be considered an example, including any mineral, vegetation, or air or water. If it can be mined, grown, caught, or collected and isn't produced by man, it's a raw material.

When a feedstock is a renewable biological substance, examples include crops, woody plants, algae, petroleum, and natural gas. Specifically, crude oil is a feedstock for the production of gasoline. In the chemical industry, petroleum is a feedstock for a host of chemicals, including methane, propylene, and butane. Algae is a feedstock for hydrocarbon fuels, Corn is a feedstock for ethanol.

Energy Crop

Energy crops are low-cost and low-maintenance crops grown solely for energy production (not for food). The crops are processed into solid, liquid or gaseous fuels, such as pellets, bioethanol or biogas. The fuels are burned to generate power or heat.

The plants are generally categorized as woody or herbaceous. Woody plants include willow and poplar, herbaceous plants include *Miscanthus x giganteus* and *Pennisetum purpureum* (both known as elephant grass). Herbaceous crops, while physically smaller than trees, store roughly twice the amount of CO_2 (in the form of carbon) below ground, compared to woody crops.

Through biotechnological procedures such as genetic modification plants can be manipulated to create higher yields. Relatively high yields can also be realized with existing cultivars. However, some additional advantages such as reduced associated costs (i.e. costs during the manufacturing process) and less water use can only be accomplished by using genetically modified crops.

CO_2 Neutrality

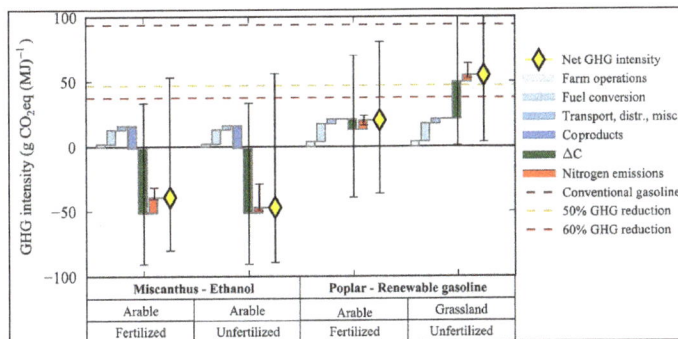

In the above figure is shown GHG/CO_2/carbon negativity for Miscanthus x giganteus production pathways.Relationship between above-ground yield (diagonal lines), soil organic carbon (X axis), and soil's potential for successful/unsuccessful carbon sequestration (Y axis). Basically, the higher the yield, the more land is usable as a GHG mitigation tool (including relatively carbon rich land.)

The amount of carbon sequestrated and the amount of GHG (greenhouse gases) emitted will determine if the total GHG life cycle cost of a bio-energy project is positive, neutral or negative. Specifically, a

GHG/carbon negative life cycle is possible if the total below-ground carbon accumulation more than compensates for the above-ground total life-cycle GHG emissions. Whitaker et al. estimates that for Miscanthus x giganteus, carbon neutrality and even negativity is within reach. Basically, the yield and related carbon sequestration is so high that it more than compensates for both farm operations emissions, fuel conversion emissions and transport emissions. The graphic on the right displays two CO_2 negative Miscanthus x giganteus production pathways, represented in gram CO_2-equivalents per megajoule. The yellow diamonds represent mean values.

One should note that successful sequestration is dependent on planting sites, as the best soils for sequestration are those that are currently low in carbon. The varied results displayed in the graph highlights this fact. Milner et al. argues that for the UK, successful sequestration is expected for arable land over most of England and Wales, with unsuccessful sequestration expected in parts of Scotland, due to already carbon rich soils (existing woodland). Also, for Scotland, the relatively lower yields in this colder climate makes CO_2 negativity harder to achieve. Soils already rich in carbon includes peatland and mature forest. Grassland can also be carbon rich, however Milner et al. further argues that the most successful carbon sequestration in the UK takes place below improved grasslands. The bottom graphic displays the estimated yield necessary to achieve CO_2 negativity for different levels of existing soil carbon saturation.

The perennial rather than annual nature of Miscanthus crops implies that the significant below-ground carbon accumulation each year is allowed to continue undisturbed. No annual plowing or digging means no increased carbon oxidation and no stimulation of the microbe populations in the soil, and therefore no accelerated carbon-to-CO_2 conversion happening in the soil every spring.

Types

Solid Biomass

Elephant grass (*Miscanthus giganteus*) is an experimental energy crop.

Solid biomass, often pelletized, is used for combustion in thermal power stations, either alone or co-fired with other fuels. Alternatively it may be used for heat or combined heat and power (CHP) production.

In short rotation coppice (SRC) agriculture, fast growing tree species like willow and poplar are grown and harvested in short cycles of three to five years. These trees grow best in wet soil conditions. An influence on local water conditions can not be excluded. Establishment close to vulnerable wetland should be avoided.

Gas Biomass (Methane)

Whole crops such as maize, Sudan grass, millet, white sweet clover, and many others can be made into silage and then converted into biogas. Anaerobic digesters or biogas plants can be directly supplemented with energy crops once they have been ensiled into silage. The fastest growing sector of German biofarming has been in the area of "Renewable Energy Crops" on nearly 500,000 ha (1,200,000 acres) of land. Energy crops can also be grown to boost gas yields where feedstocks have a low energy content, such as manures and spoiled grain. It is estimated that the energy yield presently of bioenergy crops converted via silage to methane is about 2 GWh/km² (1.8×10^{10} BTU/sq mi) annually. Small mixed cropping enterprises with animals can use a portion of their acreage to grow and convert energy crops and sustain the entire farms energy requirements with about one fifth of the acreage. In Europe and especially Germany, however, this rapid growth has occurred only with substantial government support, as in the German bonus system for renewable energy. Similar developments of integrating crop farming and bioenergy production via silage-methane have been almost entirely overlooked in N. America, where political and structural issues and a huge continued push to centralize energy production has overshadowed positive developments.

Liquid Biomass

Biodiesel

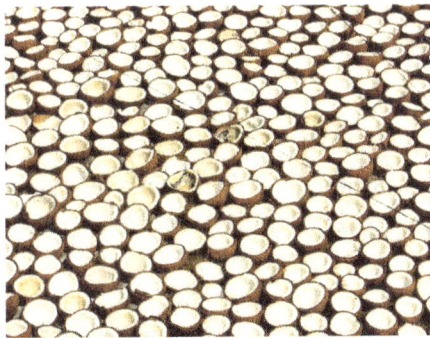

Coconuts sun-dried in Kozhikode, Kerala for making copra, the dried meat, or kernel, of the coconut. Coconut oil extracted from it has made copra an important agricultural commodity for many coconut-producing countries. It also yields coconut cake which is mainly used as feed for livestock.

Pure biodiesel (B-100), made from soybeans.

European production of biodiesel from energy crops has grown steadily in the last decade, principally focused on rapeseed used for oil and energy. Production of oil/biodiesel from rape covers more than 12,000 km² in Germany alone, and has doubled in the past 15 years. Typical yield of oil as pure biodiesel may be is 100,000 L/km² (68,000 US gal/sq mi; 57,000 imp gal/sq mi) or more, making biodiesel crops economically attractive, provided sustainable crop rotations exist that are nutrient-balanced and preventative of the spread of disease such as clubroot. Biodiesel yield of soybeans is significantly lower than that of rape.

Typical oil extractable by weight	
Crop	Oil %
copra	62
castor seed	50
sesame	50
groundnut kernel	42
jatropha	40
rapeseed	37
palm kernel	36
mustard seed	35
sunflower	32
palm fruit	20
soybean	14
cotton seed	13

Bioethanol

Energy crops for biobutanol are grasses. Two leading non-food crops for the production of cellulosic bioethanol are switchgrass and giant miscanthus. There has been a preoccupation with cellulosic bioethanol in America as the agricultural structure supporting biomethane is absent in many regions, with no credits or bonus system in place. Consequently, a lot of private money and investor hopes are being pinned on marketable and patentable innovations in enzyme hydrolysis and the like.

Bioethanol also refers to the technology of using principally corn (maize seed) to make ethanol directly through fermentation, a process that under certain field and process conditions can consume as much energy as is the energy value of the ethanol it produces, therefore being non-sustainable. New developments in converting grain stillage (referred to as distillers grain stillage or DGS) into biogas energy looks promising as a means to improve the poor energy ratio of this type of bioethanol process.

Camelina

Camelina is a feedstock for biodiesel, Camelina (Camelina Sativa L. Crantz), sometimes called "false flax" is a member of the mustard (brassica) family. It contains 30 to 40% oil by weight. The oil is mostly unsaturated (90%) and is high in Omega-3 fatty acids making it useful in a healthy human diet. The remaining seed meal contains about 40% protein and has been considered for use in poultry and livestock diets, but glucosinolates can pose a problem.

Production and Agronomic Information

Camelina is a yellow flowered annual, short season crop (85-100 days) growing 1-3 ft. tall, producing very small (<1/16") oval seeds in 1/4"-1/2" onion bulb-shaped pods. Camelina seeds are small (220,000 to 450,000 seeds/pound). A more detailed description can be found in the NRCS-USDA Plant Guide, which also contains a valuable list of other references on Camelina.

Production of Camelina can be a fall planted winter annual in more southern states to an early-planted spring annual in the northern U.S. Although evaluated on the east coast and as far south as Florida, it appears to perform best as a spring annual in the Pacific Northwest and northern Rocky Mountain states. Because of its cold hardiness, it can be planted in very early spring by drilling or broadcasting 3-5 lbs/A PLS shallow (1/4") into a firm seed bed. Several varieties are available, but they need to be selected relative to their adaptation and performance in the intended production area.

Camelina plants with seed pods growing at Tennessee State University research farm.

Camelina is resistant to blackleg (a disease common in Brassicas such as canola), has few insect problems, and competes well with weeds if grown at high densities (except for perennial weeds, which may be difficult to control). Because it is currently produced on limited acres and has not been evaluated extensively, few pesticides are labeled—should they be needed. Fertility requirements are similar to Rapeseed and Canola but at a more modest level. N needs appear to be about 5 lbs/100 lbs of expected yield. Harvest is similar to that of other oil mustards, but care must be taken because of the very small seed size. Camelina trials in Montana have reported reports yields of 400 to nearly 2,000 pounds/acre in dryland areas with 16 to 18 inches of precipitation. Idaho reports 1,700 to 2,200 pounds/acre in areas with 20 to 24 inches. Most other production areas have reported lower yields.

Production Challenges

Although Camelina is generally less expensive to grow compared to other oil mustard crops, its generally lower productivity on poorer soils where it is preferentially adapted, results in low economic returns at current prices.

Jatropha Curcas

Jatropha curcas is a species of flowering plant in the spurge family, Euphorbiaceae, that is native to the American tropics, most likely Mexico and Central America. It is originally native to the tropical areas of the Americas from Mexico to Argentina, and has been spread throughout the world in tropical and subtropical regions around the world, becoming naturalized or invasive in many areas. The specific epithet, "curcas", was first used by Portuguese doctor Garcia de Orta more than 400 years ago. Common names in English include physic nut, Barbados nut, poison nut, bubble bush or purging nut. In parts of Africa and areas in Asia such as India it is often known as "castor oil plant" or "hedge castor oil plant", but it is not the same as the usual castor oil plant, Ricinus communis (they are in the same family but different subfamilies).

J. curcas is a semi-evergreen shrub or small tree, reaching a height of 6 m (20 ft) or more. It is resistant to a high degree of aridity, allowing it to grow in deserts. It contains phorbol esters, which are considered toxic. However, edible (non-toxic) provenances native to Mexico also exist, known by the local population as piñón manso, xuta, chuta, aishte, among others. *J. curcas* also contains compounds such as trypsin inhibitors, phytate, saponins and a type of lectin known as curcin.

The seeds contain 27–40% oil (average: 34.4%) that can be processed to produce a high-quality biodiesel fuel, usable in a standard diesel engine. Edible (non-toxic) provenances can be used for animal feed and food.

Botanical Features

- Leaves: The leaves have significant variability in their morphology. In general, the leaves are green to pale green, alternate to subopposite, and three- to five-lobed with a spiral phyllotaxis.

- Flowers: Male and female flowers are produced on the same inflorescence, averaging 20 male flowers to each female flower, or 10 male flowers to each female flower. The inflorescence can be formed in the leaf axil. Plants occasionally present hermaphroditic flowers.

- Fruits : Fruits are produced in winter, or there may be several crops during the year if soil moisture is good and temperatures are sufficiently high. Most fruit production is concentrated from midsummer to late fall with variations in production peaks where some plants have two or three harvests and some produce continuously through the season.

- Seeds: The seeds are mature when the capsule changes from green to yellow. The seeds contain around 20% saturated fatty acids and 80% unsaturated fatty acids, and they yield 25–40% oil by weight. In addition, the seeds contain other chemical compounds, such as saccharose, raffinose, stachyose, glucose, fructose, galactose, and protein. The oil is largely made up of oleic and linoleic acids. Furthermore, the plant also contains curcasin, arachidic, myristic, palmitic, and stearic acids and curcin.

- Genome: The whole genome was sequenced by *Kazusa DNA Research Institute*, Chiba Japan in October 2010.

Cultivation

Jatropha curcas seeds.

Cultivation is uncomplicated. *Jatropha curcas* grows in tropical and subtropical regions. The plant can grow in wastelands and grows on almost any terrain, even on gravelly, sandy and saline soils. It can thrive in poor and stony soils, although new research suggests that the plant's ability to adapt to these poor soils is not as extensive as had been previously stated. Complete germination is achieved within 9 days. Adding manure during the germination has negative effects during that phase, but is favorable if applied after germination is achieved. It can be propagated by cuttings, which yields faster results than multiplication by seeds.

The flowers only develop terminally (at the end of a stem), so a good ramification (plants presenting many branches) produces the greatest amount of fruits. The plants are self-compatible. Another productivity factor is the ratio between female and male flowers within an inflorescence, more female flowers mean more fruits. *Jatropha curcas* thrives on a mere 250 mm (10 in) of rain a year, and only during its first two years does it need to be watered in the closing days of the dry season. Ploughing and planting are not needed regularly, as this shrub has a life expectancy of approximately forty years. The use of pesticides is not necessary, due to the pesticidal and fungicidal properties of the plant. It is used in rural Bengal for dhobi itch (a common fungal infection of the skin).

While *Jatropha curcas* starts yielding from 9–12 months time, the best yields are obtained only after 2–3 years time. The seed production is around 3.5 tons per hectare (seed production ranges from about 0.4 t/ha in the first year to over 5 t/ha after 3 years). If planted in hedges, the reported productivity of *Jatropha* is from 0.8 to 1.0 kg of seed per meter of live fence.

Propagation

Jatropha curcas can easily be propagated by both seed or cuttings. Some people recommend propagation by seed for establishment of long-lived plantations. When jatropha plants develop from cuttings, they produce many branches but yield fewer seeds and do not have enough time to develop their taproot, which makes them sensitive to wind erosion. The seeds exhibit orthodox storage behaviour and under normal treatment and storage will maintain viability at high percentages for eight months to a year. Propagation through seed (sexual propagation) leads to a lot of genetic variability in terms of growth, biomass, seed yield and oil content. Clonal techniques can help in overcoming these problems. Vegetative propagation has been achieved by stem cuttings, grafting, budding as well as by air layering techniques. Cuttings should be taken preferably from juvenile plants and treated with 200 micro gram per litre of IBA (rooting hormone) to ensure the highest level of rooting in stem cuttings. Cuttings strike root easily stuck in the ground without use of hormones.

Propagation of Jatropha curcas by stem cutting.

Propagation of Jatropha curcas by grafting.

Propagation of Jatropha curcas by Air layering Jatropha Crude Oil.

Processing

Seed extraction and processing generally needs specialized facilities.

Oil content varies from 28% to 30% and 80% extraction, one hectare of plantation will give 400 to 600 litres of oil if the soil is average.

The oily seeds are processed into oil, which may be used directly ("Straight Vegetable Oil") to fuel combustion engines or may be subjected to transesterification to produce biodiesel. Jatropha oil is not suitable for human consumption, as it induces strong vomiting and diarrhea.

Biofuel

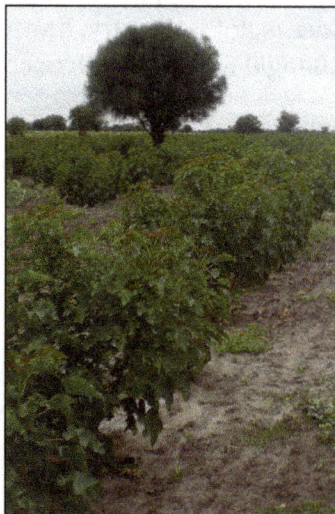

Jatropha plantation.

When jatropha seeds are crushed, the resulting jatropha oil can be processed to produce a high-quality biofuel or biodiesel that can be used in a standard diesel car or further processed into jet fuel, while the residue (press cake) can also be used as biomass feedstock to power electricity plants, used as fertilizer (it contains nitrogen, phosphorus and potassium). The cake can also be used as feed in digesters and gasifiers to produce biogas.

There are several forms of biofuel, often manufactured using sedimentation, centrifugation, and filtration. The fats and oils are turned into esters while separating the glycerin. At the end of the process, the glycerin settles and the biofuel floats. The process through which the glycerin is separated from the biodiesel is known as transesterification. Glycerin is another by-product from Jatropha oil processing that can add value to the crop. Transesterification is a simple chemical reaction that neutralizes the free fatty acids present in any fatty substances in Jatropha. A chemical exchange takes place between the alkoxy groups of an ester compound by an alcohol. Usually, methanol and ethanol are used for the purpose. The reaction occurs by the presence of a catalyst, usually sodium hydroxide (NaOH) or caustic soda and potassium hydroxide (KOH), which forms fatty esters (e.g., methyl or ethyl esters), commonly known as biodiesel. It takes approximately 10% of methyl alcohol by weight of the fatty substance to start the transesterification process.

Estimates of *Jatropha* seed yield vary widely, due to a lack of research data, the genetic diversity of the crop, the range of environments in which it is grown, and *Jatropha*'s perennial life cycle. Seed yields under cultivation can range from 1,500 to 2,000 kilograms per hectare, corresponding to extractable oil yields of 540 to 680 litres per hectare (58 to 73 US gallons per acre). In 2009 *Time* magazine cited the potential for as much as 1,600 gallons of diesel fuel per acre per year. The plant may yield more than four times as much fuel per hectare as soybean, and more than ten times that of maize (corn), but at the same time it requires five times as much water per unit of energy produced as does corn. A hectare of jatropha has been claimed to produce 1,892 litres of fuel. However, as it has not yet been domesticated or improved by plant breeders, yields are variable.

Jatropha can also be intercropped with other cash crops such as coffee, sugar, fruits and vegetables.

In 2007 Goldman Sachs cited *Jatropha curcas* as one of the best candidates for future biodiesel production. However, despite its abundance and use as an oil and reclamation plant, none of the *Jatropha* species has been properly domesticated and, as a result, its productivity is variable, and the long-term impact of its large-scale use on soil quality and the environment is unknown.

In 2008 researchers at Daimler Chrysler Research explored the use of jatropha oil for automotive use, concluding that although jatropha oil as fuel "has not yet reached optimal quality, it already fulfills the EU norm for biodiesel quality". Archer Daniels Midland Company, Bayer CropScience and Daimler AG have a joint project to develop jatropha as a biofuel. Three Mercedes cars powered by Jatropha diesel have already put some 30,000 kilometres behind them. The project is supported by DaimlerChrysler and by the German Association for Investment and Development.

Jet Fuel

Aviation fuels may be more widely replaced by biofuels such as jatropha oil than fuels for other forms of transportation. There are fewer planes than cars or trucks and far fewer jet fueling stations to convert than gas stations. To fulfil the yearly demand for aviation fuel, based on demand in 2008 (fuel use has since grown), an area of farmland twice the size of France would need to be planted with jatropha, based on average yields of mature plantations on reasonably good, irrigated land.

On December 30, 2008, Air New Zealand flew the first successful test flight from Auckland with a Boeing 747 running one of its four Rolls-Royce engines on a 50:50 blend of jatropha oil and jet A-1 fuel. In the same press release, Air New Zealand announced plans to use the new fuel for 10% of its needs by 2013. At the time of this test, jatropha oil was much cheaper than crude oil, costing an estimated $43 a barrel or about one-third of the June 4, 2008 closing price of $122.30 for a barrel of crude oil.

On January 7, 2009 Continental Airlines successfully completed a test flight from Houston, Texas using a 50/50 mixture of algae/jatropha-oil-derived biofuel and Jet A in one of the two CFM56 engines of a Boeing 737-800 Next Generation jet. The two-hour test flight could mark another promising step for the airline industry to find cheaper and more environmentally friendly alternatives to fossil fuel.

On April 1, 2011 Interjet completed the first Mexican aviation biofuels test flight on an Airbus A320. The fuel was a 70:30 traditional jet fuel biojet blend produced from Jatropha oil provided by three Mexican producers, Global Energías Renovables (a wholly owned subsidiary of U.S.-based Global Clean Energy Holdings), Bencafser S.A. and Energy JH S.A. Honeywell's UOP processed the oil into Bio-SPK (Synthetic Paraffinic Kerosene). Global Energías Renovables operates the largest Jatropha farm in the Americas.

On October 28, 2011 Air China completed the first successful demonstration flight by a Chinese airline that used jatropha-based biofuel. The mixture was a 50:50 mix of conventional jet fuel blended with jatropha oil from China National Petroleum Corp. The 747-400 powered one of its four engines on the fuel mixture during the 1-hour flight around Beijing airport.

On August 27, 2018 SpiceJet completed the first successful test flight by an Indian airline which used jatorpha based biofuel. The ratio of conventional jet fuel to jatropha oil was 25:75.

Carbon Dioxide Sequestration

According to a 2013 study published by the European Geosciences Union, the jatropha tree may have applications in the absorption of carbon dioxide, whose sequestration is important in combating climate change. This small tree is very resistant to aridity so it can be planted in hot and dry land in soil unsuitable for food production. The plant does need water to grow though, so coastal areas where desalinated seawater can be made available are ideal.

Use in Developing World

Currently the oil from *Jatropha curcas* seeds is used for making biodiesel fuel in Philippines, Pakistan and in Brazil, where it grows naturally and in plantations in the southeast, north, and northeast of Brazil. Likewise, jatropha oil is being promoted as an easily grown biofuel crop in hundreds of projects throughout India and other developing countries. Large plantings and nurseries have been undertaken in India by many research institutions, and by women's self-help groups who use a system of microcredit to ease poverty among semiliterate Indian women. The railway line between Mumbai and Delhi is planted with *jatropha* and the train itself runs on 15–20% biodiesel. In Africa, cultivation of *jatropha* is being promoted and it is grown successfully in countries such as Mali. In the Gran Chaco of Paraguay, where a native variety (*Jatropha matacensis*) also grows, studies have shown the suitability of Jatropha cultivation and agro producers are starting to consider planting in the region.

Myanmar

Myanmar is also actively pursuing the use of jatropha oil. On 15 December 2005, then-head of state, Senior General Than Shwe, said "the States and Divisions concerned are to put 50,000 acres (200 km²) under the physic nut plants [Jatropha] each within three years totalling 700,000 acres (2,800 km²) during the period". On the occasion of Burma's Peasant Day 2006, Than Shwe described in his a message that "For energy sector which is an essential role in transforming industrial agriculture system, the Government is encouraging for cultivation of physic nut plants nationwide and the technical know how that can refine physic nuts to biodiesel has also identified." He would like to urge peasants to cultivate physic nut plants on a commercial scale with major aims for emergence of industrial agriculture system, for fulfilling rural electricity supply and energy needs, for supporting rural areas development and import substitute economy.

In 2006, the chief research officer at state-run Myanma Oil and Gas Enterprise said Burma hoped to completely replace the country's oil imports of 40,000 barrels a day with home-brewed, jatropha-derived biofuel. Other government officials declared Burma would soon start exporting jatropha oil. Despite the military's efforts, the jatropha campaign apparently has largely flopped in its goal of making Burma self-sufficient in fuel.

Z.G.S. Bioenergy has started Jatropha Plantation Projects in Northern Shan State, the company has begun planting Jatropha plants during late June 2007 and will start producing seeds by 2010.

Controversies

As of 2011 scepticism about the "miracle" properties of *Jatropha* has been voiced. For example: "The idea that jatropha can be grown on marginal land is a red herring", according to Harry Stourton, former business development director of UK-based Sun Biofuels, which attempted to cultivate *Jatropha* in Mozambique and Tanzania. "It does grow on marginal land, but if you use marginal land you'll get marginal yields," he said. Sun Biofuels, after failing to adequately compensate local farmers for the land acquired for their plantation in Tanzania, pay workers severance, or deliver promised supplies to local villagers, went bankrupt later in 2011, the villager farmland being sold to an offshore investment fund.

An August 2010 article warned about the actual utility and potential dangers of reliance on *Jatropha* in Kenya. Major concerns included its invasiveness, which could disrupt local biodiversity, as well as damage to water catchment areas.

Jatropha curcas is lauded as being sustainable, and that its production would not compete with food production, but the jatropha plant needs water like every other crop to grow. This could create competition for water between the jatropha and other edible food crops. In fact, jatropha requires five times more water per unit of energy than sugarcane and corn.

Lignocellulosic Biomass

Lignocellulose refers to plant dry matter (biomass), so called lignocellulosic biomass. It is the most abundantly available raw material on the Earth for the production of biofuels, mainly bio-ethanol. It is composed of carbohydrate polymers (cellulose, hemicellulose), and an aromatic polymer (lignin). These carbohydrate polymers contain different sugar monomers (six and five carbon sugars) and they are tightly bound to lignin. Lignocellulosic biomass can be broadly classified into virgin biomass, waste biomass and energy crops. Virgin biomass includes all naturally occurring terrestrial plants such as trees, bushes and grass. Waste biomass is produced as a low value byproduct of various industrial sectors such as agriculture (corn stover, sugarcane bagasse, straw etc.) and forestry (saw mill and paper mill discards). Energy crops are crops with high yield of lignocellulosic biomass produced to serve as a raw material for production of second generation biofuel; examples include switch grass(*Panicum virgatum*) and Elephant grass.

Dedicated Energy Crops

Many crops are of interest for their ability to provide high yields of biomass and can be harvested multiple times each year. These include poplar trees and *Miscanthus giganteus*. The premier energy crop is sugarcane, which is a source of the readily fermentable sucrose and the lignocellulosic by-product bagasse.

Application
Pulp and Paper Industry

Lignocellulosic biomass is the feedstock for the pulp and paper industry. This energy-intensive industry focuses on the separation of the lignin and cellulosic fractions of the biomass.

Biofuels

Lignocellulosic biomass, in the form of wood fuel, has a long history as a source of energy. Since the middle of the 20th century, the interest of biomass as a precursor to *liquid* fuels has increased.

To be specific, the fermentation of lignocellulosic biomass to ethanol is an attractive route to fuels that supplements the fossil fuels. Biomass can be a carbon-neutral source of energy in the long run. However depending on the source of biomass, it will not be carbon neutral in the short term. For instance if the biomass is derived from trees, the time period to regrow the tree (on the order of decades) will see a net increase in carbon dioxide in the earth's atmosphere upon the combustion of lignocellulosic ethanol. However, if woody material from annual crop residue is used, the fuel could be considered carbon-neutral. Aside from ethanol, many other lignocellulose-derived fuels are of potential interest, including butanol, dimethylfuran, and gamma-Valerolactone.

One barrier to the production of ethanol from biomass is that the sugars necessary for fermentation are trapped inside the lignocellulose. Lignocellulose has evolved to resist degradation and to confer hydrolytic stability and structural robustness to the cell walls of the plants. This robustness or "recalcitrance" is attributable to the crosslinking between the polysaccharides (cellulose and hemicellulose) and the lignin via ester and ether linkages. Ester linkages arise between oxidized sugars, the uronic acids, and the phenols and phenylpropanols functionalities of the lignin. To extract the fermentable sugars, one must first disconnect the celluloses from the lignin, and then use acid or enzymatic methods to hydrolyze the newly freed celluloses to break them down into simple monosaccharides. Another challenge to biomass fermentation is the high percentage of pentoses in the hemicellulose, such as xylose, or wood sugar. Unlike hexoses such as glucose, pentoses are difficult to ferment. The problems presented by the lignin and hemicellulose fractions are the foci of much contemporary research.

A large sector of research into the exploitation of lignocellulosic biomass as a feedstock for bio-ethanol focuses particularly on the fungus *Trichoderma reesei*, known for its cellulolytic abilities. Multiple avenues are being explored including the design of an optimised cocktail of cellulases and hemicellulases isolated from *T. reesei*, as well as genetic-engineering-based strain improvement to allow the fungus to simply be placed in the presence of lignocellulosic biomass and break down the matter into D-glucose monomers. Strain improvement methods have led to strains capable of producing significantly more cellulases than the original QM6a isolate; certain industrial strains are known to produce up to 100g of cellulase per litre of fungus thus allowing for maximal extraction of sugars from lignocellulosic biomass. These sugars can then be fermented, leading to bio-ethanol.

Switchgrass

Switchgrass has excellent potential as a bioenergy feedstock for cellulosic ethanol production, direct combustion for heat and electrical generation, gasification, and pyrolysis. Switchgrass has several characteristics that make it a desirable biomass energy crop: it is a broadly adapted native to North America, it has consistently high yield relative to other species in varied environments, it requires minimal agricultural inputs, it is relatively easy to establish from seed, and a seed industry already exists.

Biology and Adaptation

Switchgrass is a perennial warm-season (C4) grass that is native to most of North America except for areas west of the Rocky Mountains and north of 55 °N latitude. Switchgrass grows 3 to 10 feet tall, typically as a bunchgrass, but the short rhizomes can form a sod over time. Switchgrass has high yield potential on marginal cropland and will be productive in most rain-fed production systems east of the 100th meridian. Productive switchgrass stands can be grown west of the 100th meridian with irrigation. Switchgrass is adapted to a wide range of habitats and climates and has few major insect or disease pests. Root depth of established switchgrass may reach 10 feet, but most of the root mass is in the top 12 inches of the soil profile. In addition to potential bioenergy production, switchgrass uses include pasture and hay production, soil and water conservation, carbon sequestration, and wildlife habitat.

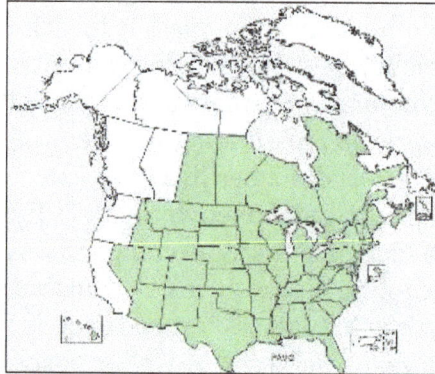

Switchgrass is adapted to much of North America.

Lowland vs Upland Ecotypes

Switchgrass has distinct lowland and upland ecotypes. Upland ecotypes occur in upland areas that are not subject to flooding, whereas lowland ecotypes are found on floodplains and other areas that receive run-on water. Generally, lowland plants have a later heading date and are taller with larger and thicker stems. Upland ecotypes are either octaploids or tetraploids, whereas lowland ecotypes are tetraploids. Lowland and upland tetraploids have been crossed to produce true F1 hybrids that have a 30 to 50% yield increase over the parental lines. These hybrids are promising sources for high-yielding bioenergy cultivars.

Production and Agronomic Information

Soybean stubble provides an excellent seedbed for no-till seeding switchgrass. During the estab-lishment year, all harvests must occur after a killing frost to avoid damaging stands. In the establishment year, good weed management and rainfall will provide about half of the fully established yield potential of the site and cultivar.

Establishing Stands

Successful stand establishment during the seeding year is mandatory for economically viable switchgrass bioenergy production systems. Weed competition is the major reason for switchgrass stand failure. Acceptable switchgrass production can be delayed by one or more years by weed competition and poor stand establishment. reported a stand frequency of 50% or greater (two or more switchgrass plants per square foot) indicated a successful stand, whereas stand frequency from 25 to 50% was marginal to adequate, and stands with less than 25% frequency indicated a partial stand. In a study conducted on 12 farms in Nebraska, South Dakota, and North Dakota, switchgrass fields with a stand frequency of 40% or greater provided a successful stand.

Switchgrass is readily established when quality seed of an adapted cultivar is used with the proper planting date, seeding rate, seeding method, and weed control. In the central Great Plains, switchgrass can be planted two or three weeks before to two or three weeks after the recommended planting dates for corn (Zea mays), typically from late April to early June. Switchgrass should be seeded at 30 pure live seed (PLS) per square foot (5 PLS pounds per acre) based on the quality of the seedlot. Excellent results are obtained by planting after a soybean (Glycine max) crop using a properly calibrated no-till drill with depth bands that plant seeds 0.25 inch to 0.5 inch deep followed by press wheels (Figure). Row spacing for switchgrass is typically 7.5 to 10 inches. If switchgrass is planted after crops that leave heavy residue such as corn or sorghum (Sorghum bicolor), it may be necessary to graze the residue, shred or bale the stalks, or use tillage to reduce the residue. If tillage is required, the seedbed needs to be packed to firm the soil. The packed soil needs to be firm enough so that walking across the field leaves only a faint footprint (Figure). Applying 8 oz of quinclorac plus 1 qt of atrazine per acre immediately after planting has provided effective grassy and broadleaf weed control for establishment. The most cost-effective method to control broadleaf weeds in switchgrass fields during the establishment year is to apply 2,4-D at 1 to 2 qt acre^{-1} after switchgrass seedlings have about four leaves. After the establishment year, a successfully established switchgrass stand requires limited herbicide applications.

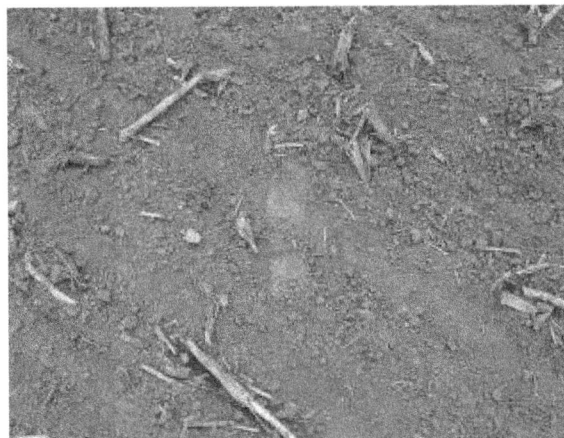

Seeding into corn or sorghum stubble may require plowing, disking, and packing to develop a firm seedbed. Pack the tilled soil until walking across the field leaves only a faint footprint to ensure good seed-to-soil contact and prevent soil in-filling of the packer wheel depression.

Nitrogen (N) fertilizer is not recommended during the planting year since N will encourage weed growth, increase competition for establishing seedlings, increase establishment cost, and increase economic risk associated with establishment if stands should fail. Soil tests are recommended prior to planting. Since switchgrass is deep rooted, soil samples should be taken from each 1-foot increment to a depth of 5 feet. In most agricultural fields, adequate levels of phosphorus (P) and potassium (K) will be in the soil profile. If warranted by soil tests, P and K can be applied before seeding to encourage root growth and promote rapid establishment. Recommended P levels for the western corn belt are in table. Switchgrass can tolerate moderately acidic soils, but optimum seed germination occurs when soil pH is between 6 and 8. With good weed management and favorable precipitation, a crop equal to about half of potential production can be harvested after frost at the end of the planting year, with 75 to 100% of full production achieved the year after planting.

Table: Phosphorus (P) recommendations for the western corn belt based on two common soil test levels.

	Soil Test Levels		
P Index Value	Bray & Kurtz #1	Olsen P (Na HCO$_3$)	P Rate
	----- ppm -----		lb P$_2$O$_5$/Acre
Very low	0-5	0-3	40
Low	6-15	4-7	20
Medium	16-25	8-14	10
High	25+	15+	0

Established Stands

Although switchgrass can survive on low fertility soils, it does respond to fertilizer, especially N. The amount of N required by switchgrass is a function of the yield potential of the site, productivity of the cultivar, and other management practices being used. Consequently, the optimum N rate for switchgrass managed for biomass will vary, but a few references indicative of the responses to N in different regions of the United States are included. Additionally, biomass will decline over years if inadequate N is applied, and yield will be sustainable only with proper N application. In Nebraska and Iowa, Cave-in-Rock yield increased as N rate increased from 0 to 270 lb N acre[-1], but soil N increased when more than 100 lb N acre[-1] were applied. They reported biomass was optimized by applying 100 lb N acre[-1], with about the same amount of N being applied as was being removed by the crop. A general N fertilizer recommendation for the Great Plains and Midwest region is to apply 20 lb N acre-1 yr[-1] for each ton of anticipated biomass if harvesting during the growing season, with N rate reduced to 12 to 14 lb N acre-1 yr[-1] for each ton of anticipated biomass if harvesting after a killing frost. The N rate can be reduced when the harvest is after a killing frost because switchgrass cycles some N back to roots during autumn. If soil tests indicate a new switchgrass field has high residual N levels, N rates can be significantly reduced during the initial production years using the above information as a guideline. Apply N at switchgrass green-up to minimize cool-season weed competition.

Table: Switchgrass publications addressing nitrogen fertilizer application for different regions of the United States listed by state and the major parameters evaluated in the study.

State(s)	Parameters Evaluated
AL	N rate and row spacing effect on C partitioning.
IA	Yield and quality parameters for 20 strains.
IA, NE	Harvest date and N rate effects.
NC, KY, TN, VA, WV	Long-term yield under different management regimes.
SD	Harvest date and N rate effects on biomass, persistence, species composition, and soil organic carbon of switchgrass-dominated CRP.
TX	Yield and stand responses to N and P as affected by row spacing.

Spraying herbicides to control broadleaf weeds typically is needed only once or twice every 10 years in an established, well-managed switchgrass stands. When needed, the most effective and economical approach is with broadcast applications of 2,4-D at 1 to 2 qt acre^{-1}. Spray broadleaf weeds as early in the growing season as possible to reduce the impact of weed interference on switchgrass yield. In some cases, cool-season grasses may invade switchgrass stands and reduce yield. Harvesting after switchgrass senescence in autumn but while cool-season grasses are growing, then applying glyphosate at 1 to 2 qt acre^{-1}, is an effective method to reduce cool-season grasses. However, make certain switchgrass is dormant when glyphosate is applied, or stands could be damaged. Spring applications of atrazine at 2 qt acre^{-1} can be used to control cool-season grasses in established switchgrass stands.

Rotary head mowers (disc mowers) effectively harvested this 6-ton per acre switchgrass field at anthesis. Additionally, after a killing frost, the multidirectional arrangement of the switchgrass in the windrow was easier to bale than the linearly arranged windrow left by the sickle-bar head.

Harvest and Storage

Maximizing yield currently is the primary objective when harvesting biomass feedstocks. In the Great Plains and Midwest, maximum first-cut yields are attained by harvesting switchgrass when panicles are fully emerged to the post-anthesis stage (~1 August). Sufficient regrowth may occur about one year out of four to warrant a second harvest after a killing frost. Do not harvest switchgrass within six weeks of the first killing frost or shorter than a 4-inch stubble height to ensure translocation of storage carbohydrates to maintain stand productivity and persistence. Dormant season harvests after a killing frost will not damage switchgrass stands but will reduce the amount of snow captured during winter. In general, a single harvest during the growing season maximizes switchgrass biomass recovery, but harvesting after a killing frost will ensure stand productivity and persistence, especially when drought conditions occur, and reduce N fertilizer requirements. Delaying harvest until spring will reduce moisture and ash contents, but yield loss can be as high as 40% compared with a fall harvest. With proper management, productive stands can be maintained indefinitely and certainly for more than 10 years. Harvesting switchgrass in summer at or after flowering when drought conditions exist is not recommended.

Switchgrass can be harvested and baled with commercially available haying equipment. Self-propelled harvesters equipped with a rotary head (disc mowers) have most effectively harvested high-yielding (>6-ton per acre) switchgrass fields. Additionally, after a killing frost, the multidirectional arrangement of the switchgrass in the windrow was easier to bale than the linearly arranged windrow left by a sickle-bar head. Round bales tend to have less storage losses than large square bales (>800 lb) when stored outside, but square bales tend to be easier to handle and load a truck for transport without road width restrictions.

After harvest, poor switchgrass storage conditions can result in storage losses of 25% in a single year. In addition to storage losses in weight, there can be significant reductions in biomass quality, and the biomass may not be in acceptable condition for a biorefinery. Switchgrass grown for use in a biorefinery may have to be stored for a full year or longer since biorefineries will operate 365 days a year. Some type of covered storage will be necessary to protect the producer's investment.

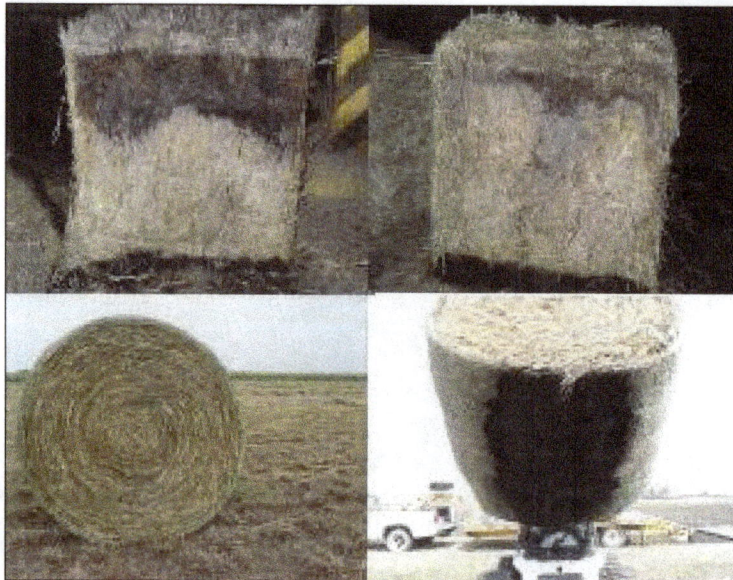

In figure Proper storage of switchgrass bales is imperative to maintain total harvested dry matter and prevent spoilage. Large square bales can spoil from the top and bottom and can lose more than 25% of total dry matter in six months when stored outside in the open (top left), but covering the large square bales with hay tarps (top right) reduces dry matter loss to about 7% in six months. Wrapping big round bales with at least three wraps of net-wrap maintains the structure of the bale and reduces the surface area of the bale that contacts the ground. Covering big round bales stored outside can reduce dry matter loss to less than 3% in six months.

Potential Yield

Switchgrass yield is strongly influenced by precipitation, fertility, soil, location, genetics, and other factors. Most plot and field-scale switchgrass research has been conducted on forage-type cultivars selected for other livestock-based characteristics in addition to yield. Consequently, the forage-type cultivars in the Great Plains and Midwest are entirely represented by upland ecotypes which are inherently lower yielding than lowland ecotypes. Thus, yield data comparing forage-type upland cultivars like Cave-In-Rock, Shawnee, Summer, and Trailblazer do not capture the full yield potential of switchgrass and are not fair comparisons. For example in Nebraska, high-yielding F1 hybrids of Kanlow and Summer produced 9.4 tons acre-1 year-1, which was 68% greater than Summer and 50% greater than Shawnee. New biomass-type switchgrass cultivars will be available in the near future for the Great Plains and Midwest. Knowing the origin of a switchgrass cultivar is important since switchgrass is photoperiod sensitive. Planting a switchgrass cultivar too far north of the cultivar origin area (>300 miles) can result in winter stand loss. Planting a switchgrass cultivar south of its origin area results in less biomass because the shorter photoperiod causes plants to flower too early.

Production Challenges

There are major challenges to using switchgrass for cellulosic ethanol. An ethanol plant requires a reliable and consistent feedstock supply. A 50-million-gallon per year plant will require 625,000 U.S. tons of feedstock per year assuming 80 gallons of ethanol can be produced from one ton of feedstock. Although cellulosic ethanol plants likely will use multiple feedstocks, this example assumes switchgrass will be the only feedstock. Operating every day of the year, the plant will require 1,712 dry matter (DM) tons of feedstock per day, or 342 acres of switchgrass yielding 5 DM tons per acre. If a loaded semi can deliver 30 round bales each containing 0.6 DM tons (18 U.S. tons), the ethanol plant will use 95 semi loads of feedstock per day, requiring a semi to be unloaded every 15 minutes 24 hours per day, 7 days per week.

There must be an available land base in the local agricultural landscape to produce feedstock. The biomass and ethanol yield of the feedstock will determine the land area required for feedstock production. Assuming 25 miles is the maximum economically feasible distance feedstock can be transported, all of the feedstock must be grown within a 25-mile radius of the bio-refinery, an area containing about 1.26 million acres. Assuming a 50-million-gallon per year cellulosic ethanol plant requires 625,000 tons of feedstock per year, if feedstock yield is 1.75 DM tons/acre, 28% of the land would need to grow the feedstock, and this is not feasible in most agricultural areas. At 5 DM tons/acre, a commonly-achieved yield with available forage cultivars, only 10% of the land would be needed for feedstock production and is feasible in most agricultural areas. However, at 10 DM tons/acre, only 5% of the land base would be needed for feedstock production and would minimally alter the agricultural landscape. Dry matter yield will exceed 10 tons/acre in many areas of the South and Southeast, so less than 5% of the land base would be needed for feedstock production. This example reinforces the importance of high DM yield to the agricultural feasibility of cellulosic ethanol, not to mention the inability of the producer to profit by growing low-yielding energy crops. A majority of the switchgrass likely will be grown on marginal lands that have suboptimal characteristics (i.e., slope, soil depth, etc.) for producing food and feed, or on lands currently enrolled in conservation programs.

In table is reported dry matter (DM) yield, acres required to grow 625,000 tons of dry matter per year, and the percent of the land base required to provide feedstock for a 50-million-gallon cellulosic ethanol plant for different herbaceous perennial feedstocks in the Great Plains and Midwest.

Feedstock	Yield, DM tons/acre	Acres needed to grow 625,000 DM tons/year	Percent of land in 25-mile radius
LIHD prairie	1.75	357,000	28
Managed native prairie	2.5	250,000	20
Shawnee switchgrass	5	125,000	10
Bioenergy switchgrass	7.4	84,460	6.6
Hybrid switchgrass	9.4	66,489	5.3

Growing switchgrass must be profitable for the producer, it must fit into existing farming operations, it must be easy to store and deliver to the ethanol plant, and extension efforts must be provided to inform producers on the agronomics and best management practices for specific regions, all of which have been addressed for switchgrass. Switchgrass fits well into the production systems of most farmers. Harvesting switchgrass after frost is a time when most farmers have completed corn and soybean harvests and handling switchgrass as a hay crop is not foreign to most producers. The economic opportunities of switchgrass for small, difficult-to-farm, or poorly-productive fields will be attractive to many producers.

There are potential difficulties with large-scale switchgrass monocultures, but most are speculation at this point. Concerns arise for potential disease and insect pests, and the escape of switchgrass as an invasive species with the production of millions of switchgrass acres, especially since little research has been conducted on these topics. Most pathogen issues cannot be fully realized until large areas are planted to switchgrass. However, the broad genetic diversity available to switchgrass breeders, the initial pathogen screening conducted during cultivar development, and the fact that switchgrass has been a native component of central U.S. grasslands for centuries will likely limit the negative pest issues. Switchgrass has been used widely throughout the Great Plains and Midwest for pasture and conservation purposes for decades, and no invasive problems have developed or been identified.

Production Cost

Results of a recent economic study based on the five-year average of 10 farms in Nebraska, South Dakota, and North Dakota indicated producers can grow switchgrass at a farm gate cost of $60/ton. However, producers with experience growing switchgrass had five-year average costs of $43/ton, and one producer grew switchgrass for $38/ton. These costs include all expenses plus land costs and labor at $10/hour. Each big round bale represents 50 gallons of ethanol assuming 80 gallons per ton of switchgrass, with a farm gate cost of $0.75/gallon at $60/ton. This research from nearly 50 production environments indicates that growing switchgrass for cellulosic ethanol is economically feasible in the central and northern Great Plains. It should be noted that fuel and land prices have increased since this study, so the cost increases for those inputs need to be considered when determining switchgrass production costs.

Environmental and Sustainability Issues

Sustainable biomass energy crops must be productive, protective of soil and water resources, and profitable for the producer. Numerous studies have reported that switchgrass will protect soil, water, and air quality; provide fully sustainable production systems; sequester C; create wildlife habitat; increase landscape and biological diversity; return marginal farmland to production; and increase farm revenues. Switchgrass root density in the surface 6 inches is two-fold greater than alfalfa, more than three-fold greater than corn, and more than an order of magnitude greater than soybean. In a five-year field study conducted on 10 farms in Nebraska, South Dakota, and North Dakota, Liebig et al. reported that switchgrass stored large quantities of C, with four farms in Nebraska storing an average of 2,590 pounds of soil organic C (SOC) acre-1 year-1 when measured to a depth of 4 feet. However, they noted that SOC increases varied across sites, and the variation in SOC change reiterated the importance of long-term environmental monitoring sites in major agro-ecoregions.

Energy produced from renewable carbon sources is held to a different standard than energy produced from fossil fuels, in that renewable fuels must have highly positive energy values and low greenhouse gas emissions. The energy efficiency and sustainability of ethanol produced from grains and cellulosics has been evaluated using net energy value (NEV), net energy yield (NEY), and the ratio of the biofuel output to petroleum input [petroleum energy ratio (PER)]. An energy model using estimated agricultural inputs and simulated yields predicted switchgrass could produce greater than 700% more output than input energy. These modeled results were validated with actual inputs from multi-farm, field-scale research to predict energy output. Switchgrass fields on 10 farms in Nebraska, South Dakota, and North Dakota produced 540% more renewable energy (NEV) than nonrenewable energy consumed over a five-year period (Schmer et al., 2008). The estimated on-farm NEY was 93% greater than human-made prairies and 652% greater than low-input switchgrass grown in small plots in Minnesota. The 10 farms and five production years had a PER of 13.1 and produced 93% more ethanol per acre than human-made prairies and 471% more ethanol per acre than low-input switchgrass in Minnesota. Average greenhouse gas (GHG) emissions from switchgrass-based ethanol were 94% lower than estimated GHG emissions for gasoline. Switchgrass for bioenergy is an energetically positive and environmentally sustainable production system for the Great Plains.

Implementing switchgrass-based bioenergy production systems will require converting marginal land from annual row crops to switchgrass and could exceed 10% in some regions depending on the yield potential of the switchgrass strains. In a five-year study in Nebraska, the potential ethanol yield of switchgrass averaged 372 gallons acre-1 and was equal to or greater than that for no-till corn (grain + stover) on a dry-land site with marginal soils. Removing 50% of the corn stover each year reduced subsequent corn grain yield, stover yield, and total biomass. Growing switchgrass on marginal sites likely will enhance ecosystem services more rapidly and significantly than on more productive sites.

Feasibility

Perennial herbaceous energy crops provide several challenges. A stable and consistent feedstock supply must be available year-round to the ethanol or power plant. For the producer, perennial herbaceous energy crops must be profitable, they must fit into existing farming operations, they must be easy to store and deliver to the plant, and extension efforts must be provided to inform producers on the agronomics and best management practices for growing perennial herbaceous energy crops. However, perennial herbaceous energy crops have potential for improvement, and they present a unique opportunity for cultural change on the agricultural landscape. There are numerous environmental benefits to perennial herbaceous cropping systems that can improve agricultural land use practices such as stabilizing soils and reducing soil erosion, improved water quality, increased and improved wildlife habitat, and storing C to mitigate greenhouse gas emissions. There is large potential for achieving all of these benefits, provided agronomic, genomic, and operational aspects of perennial herbaceous cropping systems are fully developed and accepted by farmers. Herbaceous perennial energy crops may be used in conjunction with agriculture residues (corn stover and wheat straw), which likely would be harvested in autumn, and perennial grasses could be harvested in very early spring while they are dry, similar to when prairies are typically burned. This may help reduce the need for feedstock storage by providing feedstock at different times during the year.

Growing seed to meet potential demand for bioenergy will not be an issue. Switchgrass has many desirable seed characteristics and can produce viable seed during the seeding year, especially under irrigation. Established seed production fields can produce 500 to 1,000 pounds of seed per acre with irrigation, and the seed is easily threshed, cleaned, and planted with commercial planting equipment. Seed production systems are well established, and a commercial industry for switchgrass seed has existed for over 50 years.

Miscanthus Giganteus

Miscanthus x giganteus (miscanthus giganteus, giant miscanthus, elephant grass) is a sterile hybrid of Miscanthus sinensis and Miscanthus sacchariflorus. It can grow to heights of more than 4 metres (13 ft) in one growing season (from the third season onwards). Just like Pennisetum purpureum and Saccharum ravennae it is also called elephant grass.

M. x giganteus' perennial nature, its ability to grow on marginal land, its water efficiency, non-invasiveness, low fertilizer needs, significant carbon sequestration and high yield have sparked a lot of interest among researchers, with some arguing that it has "ideal" energy crop properties. Some argue that it has the potential to be a greenhouse gas (GHG) negative fuel, while others highlight its water cleaning and soil enhancing qualities. There are practical and economic challenges related to its use in the existing, fossil based combustion infrastructure, however. Torrefaction and other fuel upgrading techniques are being explored as countermeasures to this problem.

Areas of Usage

M. x giganteus is mainly used as raw material for solid biofuels. It can be burned directly, or processed further into pellets or briquettes. It can also be used as raw material for liquid biofuels or biogas.

It is possible to use Miscanthus as a building material, and as insulation. Materials produced from Miscanthus include fiberboards, composite Miscanthus/wood particleboards, and blocks. It can be used as raw material for pulp and fibers as well as molded products such as eco-friendly disposable plates, cups, cartons, etc. Miscanthus has a pulp yield of 70-80% compared to dry weight, due to the high holocellulose content. The pulp can be processed further into methylcellulose and used as a food additive and in many industrial applications. Miscanthus fiber provides raw material for reinforcement of biocomposite or synthetic materials. In agriculture, Miscanthus straw is used in soil mulching to retain soil moisture, inhibit weed growth, and prevent erosion. Further, Miscanthus' high carbon to nitrogen ratio makes it inhospitable to many microbes, creating a clean bedding for poultry, cattle, pigs, horses, and companion animals. Miscanthus used as horse bedding can be combined with making organic fertilizer. Miscanthus can be used as a healthy fiber source in pet food.

Life Cycle

Propagation

Miscanthus x giganteus is propagated by cutting the rhizomes (its underground stems) into small pieces, and then re-planting those pieces 10 cm (4 in) underground. One hectare (2.5 acres) of Miscanthus rhizomes, cut into pieces, can be used to plant 10-30 hectares of new Miscanthus fields (multiplication factor 10-30). Rhizome propagation is a labor-intensive way of planting new crops, but only happens once during a crop's lifetime. New and cheaper propagation techniques is underway, which seem to increase the multiplication factor from 10-30 to 1000-2000. A halving of the cost is predicted.

Management

A limited amount of herbicide should only be applied at the beginning of the first two seasons; after the second year the dense canopy and the mulch formed by dead leaves effectively reduces weed growth. Other pesticides are not needed. Because of Miscanthus' high nitrogen efficiency, fertilizer is also usually not needed. Mulch film, on the other hand, helps both M. x giganteus and various seed based hybrids to grow faster and taller, with a larger number of stems per plant, effectively reducing the establishment phase from three years to two. The reason seems to be that this plastic film keeps the humidity in the topsoil and increases the temperature.

Yield – Overview

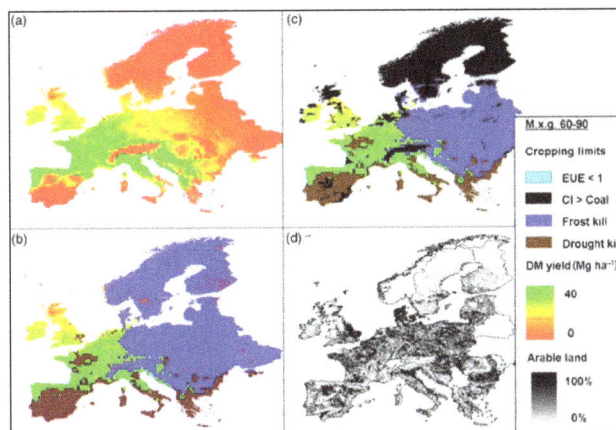

Yield estimate for Miscanthus x giganteus in Europe (no irrigation).

Miscanthus x giganteus is close to the theoretical maximum efficiency at turning solar radiation into biomass, and its water use efficiency is among the highest of any crop. It has twice the water use efficiency as its fellow C4 plant maize, and four times the efficiency as the C3 plant wheat. The efficiency has as an important consequence: It makes Miscanthus x giganteus fields energy dense. Since Miscanthus has an energy content of 18 GJ per dry tonne, the typical UK dry yield of 11-14 tonnes per hectare produce 200-250 gigajoules of energy per hectare per year (200-250 GJ ha-1 yr-1). This compares favorably to maize (98 GJ), oil seed rape (25 GJ), and wheat/sugar beet (7-15 GJ), underlining the differences between first and second generation bioenergy crops.

In Europe the peak (autumn) dry mass yield has been measured to 10–40 tonnes per hectare per year (4–16 tonnes per acre per year), depending on location, with a mean peak dry mass yield of 22 tonnes. European yields are highest in southern Europe. Trials in Illinois, USA, had yields 10–15 tonnes per acre (25–37 t/ha). Like Europe, yields decrease as you move north. M. x giganteus has been shown to yield two to three times more than switchgrass.

Peak yield is reached at the end of summer but harvest is typically delayed until winter or early spring. Yield is roughly 33% lower at this point because of leaves drop, but the combustion quality is higher. Delayed harvest also allows nitrogen to move back into the rhizome for use by the plant in the following growing season.

Yield – Arable Land

In Germany, Felten et al. did a 16-year trial on arable land and concluded with a mean winter/spring yield of 15 tonnes per hectare per year (6.1 tonnes per acre per year). McCalmont et al. estimates a mean UK yield of 10-15 tonnes (if harvested in the spring), while Hastings et al. estimates a "pessimistic" UK mean yield of 10.5 tonnes. Nsanganwimana et al. summarizes several trials, and give these numbers:

- Austria: Autumn harvest 17-30. Winter harvest 22.
- Denmark: Autumn harvest 17. Winter harvest 10.
- France: Autumn harvest 42-49. Winter harvest 30.
- Germany: Autumn harvest 17-30. Winter harvest 10-20.
- Portugal: Autumn harvest 39. Winter harvest 26-30.
- The Netherlands: Autumn harvest 25. Winter harvest 16-17.
- Spain: Winter harvest 14.
- UK: Winter harvest 11-17.

Yield – Marginal Land

Marginal land is land with issues that limits growth, for instance low water and nutrient storage capacity, high salinity, toxic elements, poor texture, shallow soil depth, poor drainage, low fertility, or steep terrain. Depending on how the term is defined, between 1.1 and 6.7 billion hectares of marginal land exists in the world. For comparison, Europe consists of roughly 1 billion hectares (10 million km2, or 3.9 million square miles), and Asia 4.5 billion hectares (45 million km2, or 17 million square miles).

Steep, marginal land.

Quinn et al. identified Miscanthus x giganteus as a crop that is moderately or highly tolerant of multiple environmental stressors, specifically, heat, drought, flooding, salinity (below 100 mM), and cool temperatures (down to −3.4 °C, or 25 °F). This robustness makes it possible to establish relatively high-yielding Miscanthus fields on marginal land, Nsanganwimana et al. mentions wastelands, coastal areas, damp habitats, grasslands, abandoned milling sites, forest edges, streamsides, foothills and mountain slopes as viable locations. Likewise, Stavridou et al. concluded that 99% of Europe's saline, marginal lands can be used for M. x giganteus planta-tions, with only an expected maximum yield loss of 11%. Since salinity up to 200 mM does not affect roots and rhizomes, carbon sequestration carry on unaffected. Lewandowski et al. found a yield loss of 36% on a marginal site limited by low temperatures (Moscow), compared to max-imum yield on arable land in central Europe. They also found a yield loss of 21% on a marginal site limited by drought (Turkey), compared to maximum yields on arable soil in central Europe. Further, Nsanganwimana et al. found that M. x giganteus grows well in soils contaminated by metals, or by industrial activities in general. For instance, in one trial, it was found that M. x giganteus absorbed 52% of the lead content and 19% of the arsenic content in the soil after three months. The absorption stabilizes the pollutants so they don't travel into the air (as dust), into ground water, neighbouring surface waters, or neighbouring areas used for food production. If contaminated Miscanthus is used as fuel, the combustion site need to install the appropriate equipment to handle this situation. On the whole though, "Miscanthus is a suitable crop for combining biomass production and ecological restoration of contaminated and marginal land." Clifton-Brown et al. concludes that Miscanthus x giganteus can "contribute to the sustainable in-tensification of agriculture, allowing farmers to diversify and provide biomass for an expanding market without compromising food security."

Carbon Sequestration

Soil Carbon Input/Output

Plants sequester carbon through photosynthesis, by exchanging O_2 (oxygen) for CO_2, thus keeping the carbon (C) to itself. When the plants move carbon to the roots, it does not stay down there forever however; "soil carbon is a balance between the decay of the initial soil carbon and the rate of input." Plant derived soil carbon is a continuum, ranging from living biomass to humus, and it decays in different stages, ranging from months (decomposable plant material; DPM) to hun-dreds of years (humus). The rate of decay depends on many factors, for instance plant species, soil, temperature and humidity, but as long as fresh new carbon is inputted, a certain amount of carbon stay in the ground – in fact Poeplau et al. did not find any "indication of decreasing SOC [soil organic carbon] accumulation with age of the plantation indicating no SOC saturation within 15–20 years." The amount of carbon in the ground under Miscanthus fields is thus seen to increase during the entire life of the crop, albeit with a slow start because of the initial tilling (plowing, digging) and the relatively low amounts of carbon input in the establishment phase. (Tilling induces soil aeration, which accelerates the soil carbon decomposition rate, by stimulating soil microbe populations. Also, tilling makes it easier for the oxygen (O) atoms in the atmosphere to attach to carbon (C) atoms in the soil, producing CO_2). Felten et al. argues that high proportions of pre- and direct-harvest residues (e.g. dead leaves), direct humus accumulation, the well-developed and deep-reaching root system, the low decomposition rates of plant residues due to a high C: N ratio (carbon to nitrogen ratio), and the absence of tillage and subsequently less soil aeration are the reasons for the high carbon sequestration rates.

At the end of each season, the plant pulls the nutrients to the ground.
The color shifts from green to yellow/brown.

Net Annual Carbon Accumulation

A number of studies try to quantify the net amount of below-ground carbon accumulation each year, after decay is accounted for, in various locations and under various circumstances.

Dondini et al. found 32 tonnes more carbon per hectare (13 tonnes per acre) under a 14 year old Miscanthus field than in the control site, suggesting a combined (C3 plus C4) accumulation rate of 2.29 tonnes per hectare (0.91 long ton/acre), or 38% of total harvested carbon per year. Likewise, Milner et al. suggest a mean carbon accumulation rate for the whole of the UK of 2.28 tonnes per hectare (0.91 long ton/acre) per year (also 38% of total harvested carbon per year), given that some unprofitable land (0.4% of total) is excluded. Nakajima et al. found an accumulation rate of 1.96 (±0.82) tonnes per hectare per year below a university test site in Sapporo, Japan (0.79 per acre), equivalent to 16% of total harvested carbon per year. The test was shorter though, only 6 years. Hansen et al. found an accumulation rate of 0.97 tonne per hectare per year (0.39 tonnes per acre per year) over 16 years under a test site in Hornum, Denmark, equivalent to 28% of total harvested carbon per year. McCalmont et al. compared a number of individual European reports, and found accumulation rates ranging from 0.42 to 3.8 tonnes per hectare per year, with a mean accumulation rate of 1.84 tonne (0.74 tonnes per acre per year), or 20% of total harvested carbon per year.

Transport and Combustion Challenges

Biomass in general, including Miscanthus x giganteus, have different properties compared to coal, for instance when it comes to handling and transport, grinding, and combustion. This makes sharing the same logistics, grinding and combustion infrastructure difficult. Often new biomass handling facilities have to be built instead, which increases cost. Together with the relatively high cost of feedstock, this often lead to the well-known situation where biomass projects has to receive subsidies to be economically viable. A number of fuel upgrading technologies are currently being explored however that makes biomass more compatible with the existing infrastructure. The most mature of these is torrefaction, basically an advanced roasting technique which – when combined with pelleting or briquetting – significantly influences both the handling and transport properties, grindability and combustion efficiency.

Energy Density and Transport Costs

Transport of bulky, water absorbing Miscanthus bales.

Miscanthus bales and chips have a bulk density of approximately 150 kg/m³, while briquettes have a bulk density of up to 600 kg/m³. Torrefaction increases bulk density further, as approximately two thirds of the original mass is retained as a solid product, while approximately one third of the original mass was converted to gas during the process. The finished, torrefied, solid product, in the form of pellets or briquettes, still contains approximately 85% of the original energy, however. Basically the mass part reduce more than the energy part, and the consequence is that the calorific value of torrefied biomass approaches that of medium grade coal - typically the calorific value increases from 18 GJ per tonne dry mass to 23 GJ per tonne torrefied mass.

The higher energy density means lower transport costs, and a decrease in transport-related GHG emittance. The IEA (International Energy Agency) has calculated energy and GHG costs for regular and torrefied pellets/briquettes. When making pellets and shipping them from Indonesia to Japan, a minimum 6.7% of energy savings or 14% GHG savings is expected when switching from regular to torrefied. This number increases to 10.3% energy savings and 33% GHG savings when making and shipping minimum 50mm briquettes instead of pellets. The longer the route, the bigger the savings. The relatively short supply route from Russia to the UK equals energy savings of 1.8%, while the longer supply route from southeast USA to the Amsterdam-Rotterdam-Antwerp (ARA) area is 7.1%. From southwest Canada to ARA 10.6%, southwest USA to Japan 11%, and Brazil to Japan 11.7% (all these savings are for pellets only).

Water Absorption and Transport Costs

Torrefaction also converts the biomass from a hydrophilic (water absorbing) to a hydrophobic (water repelling) state. Water repelling briquettes can be transported and stored outside, which simplifies the logistics operation and decreases cost. Almost all biological activity is stopped, reducing the risk of fire and biological decomposition like rotting.

Uniformity and Customization

Generally, torrefaction is seen as a gateway for converting a range of very diverse feedstocks into a uniform and therefore easier to deal with fuel. The fuel's parameters can be changed to meet customers demands, for instance durability, water resistance, ash composition, torrefaction degree, geometrical form, and type of feedstock. The possibility to use different types of feedstock improves the fuel's availability and supply reliability.

Grindability

Coal grinders.

Unprocessed M. x giganteus has strong fibers, making grinding into equally sized, very small particles (below 75 μm/0.075 mm) difficult to achieve. Coal chunks are typically ground to that size because such small, even particles combust stabler and more efficient than the larger fuel chunks. While coal has a score on the Hardgrove Grindability Index (HGI) of 30-100 (higher numbers means it is easier to grind), unprocessed Miscanthus has a score of 0. During torrefaction however, "the hemi-cellulose fraction which is responsible for the fibrous nature of biomass is degraded, thereby improving its grindability." Bridgeman et al. measured a HGI of 79 for torrefied Miscanthus, Smith et al. measured a HGI of 150 for Miscanthus pre-treated with hydrothermal carbonisation: while the IEA estimates a HGI of 23-53 for torrefied biomass in general. UK coal scores between 40 and 60 on the HGI scale. The IEA estimates an 80-90% drop in energy use required to grind biomass that has been torrefied.

The relatively easy grinding of torrefied Miscanthus makes a cost-effective conversion to fine particles possible, which subsequently makes efficient combustion with a stable flame possible. Ndibe et al. found that the level of unburnt carbon "decreased with the introduction of torrefied biomass", and that the torrefied biomass flames "were stable during 50% cofiring and for the 100% case as a result of sufficient fuel particle fineness."

Chlorine and Corrosion

Raw miscanthus biomass has a relatively high chlorine amount, which is problematic in a combustion scenario because, as Ren et al. explains, the "likelihood of corrosion depends significantly on the content of chlorine in the fuel." Likewise, Johansen et al. states that "the release of Cl-associated [chlorine-associated] species during combustion is the main cause of the induced active corrosion in the grate combustion of biomass." Chlorine in different forms, in particular combined with potassium as potassium chloride, condensates on relatively cooler surfaces inside the boiler and creates a corrosive deposit layer. The corrosion damages the boiler, and in addition the physical deposit layer itself reduce heat transfer efficiencey, most critically inside the heat exchange mechanism. Chlorine and potassium also lowers the ash melting point considerably compared to coal. Melted ash, known as slag or clinker, sticks to the bottom of the boiler, and increase maintenance costs.

In order to reduce chlorine (and moisture) content, M. x giganteus is usually harvested dry, in early spring, but this late harvest practice is still not enough of a countermeasure to achieve corrosion-free combustion.

However, Ren et al. found that "59.1 wt%, 60.7 wt% and 77.4 wt% of the chlorine contents of olive residues, DDGS and corn straw, respectively, were released during torrefaction", and concludes that chlorine emissions "were drastically lower, by 2–5 times, than those of their raw biomass precursors." Chlorine release during the torrefaction process itself is more manageble than chlorine release during combustion, because "the prevailing temperatures during the former process are below the melting and vaporization temperatures of the alkali salts of chlorine, thus minimizing their risks of slagging, fouling and corrosion in furnaces."

For potassium, Kambo et al. found a 30% reduction for torrefied miscanthus, while Ren et al. found an 86% increase for torrefied corn stover. However, potassium is dependent on chlorine to form potassium chloride; with a low level of chlorine, the potassium chloride deposits reduce proportionally.

Li et al. concludes that the "process of torrefaction transforms the chemical and physical properties of raw biomass into those similar to coal, which enables utilization with high substitution ratios of biomass in existing coal-fired boilers without any major modifications." Likewise, Bridgeman et al. states that since torrefaction removes moisture, creates a grindable, hydrophobic and solid product with an increased energy density, torrefied fuel no longer requires "separate handling facilities when co-fired with coal in existing power stations." Smith et al. makes a similar point in regard to hydrothermal carbonization, sometimes called "wet" torrefaction.

Ribeiro et al. note that "torrefaction is a more complex process than initially anticipated" and states that "torrefaction of biomass is still an experimental technology." Michael Wild, president of the International Biomass Torrefaction Council, stated in 2015 that the torrefaction sector is "in its optimisation phase", i.e. it is maturing. He mentions process integration, energy and mass efficiency, mechanical compression and product quality as the variables most important to master at this point in the sector's development.

Environmental Impacts

Yield and Soil Carbon Content

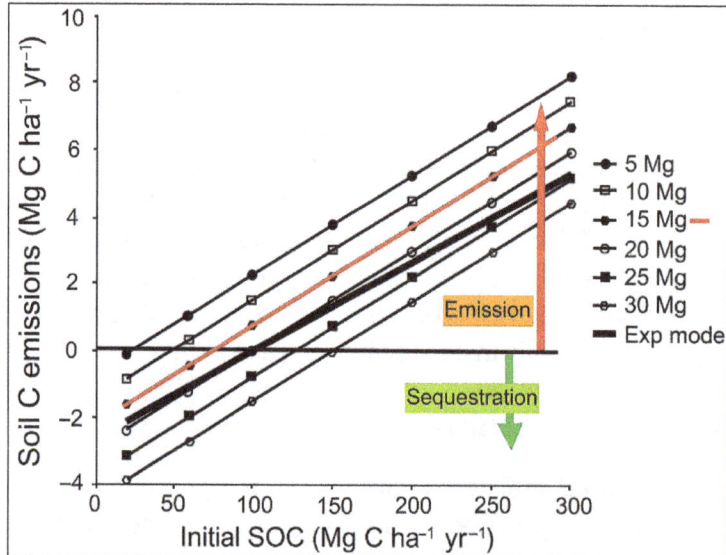

Relationship between above-ground yield (diagonal lines), soil organic carbon (X axis), and soil's potential for successful/unsuccessful carbon sequestration (Y axis). Basically, the higher the yield, the more land is usable as a GHG mitigation tool (including relatively carbon rich land).

The amount of carbon sequestrated and the amount of GHG (greenhouse gases) emitted will determine if the total GHG life cycle cost of a bio-energy project is positive, neutral or negative. Specifically, a GHG/carbon negative life cycle is possible if the total below-ground carbon accumulation more than compensates for the above-ground total life-cycle GHG emissions. Whitaker et al. estimates that for Miscanthus x giganteus, carbon neutrality and even negativity is within reach. Basically, the yield and related carbon sequestration is so high that it more than compensates for both farm operations emissions, fuel conversion emissions and transport emissions. The graphic on the right displays two CO_2 negative Miscanthus x giganteus production pathways, represented in gram CO_2-equivalents per megajoule. The yellow diamonds represent mean values.

One should note that successful sequestration is dependent on planting sites, as the best soils for sequestration are those that are currently low in carbon. The varied results displayed in the graph highlights this fact. Milner et al. argues that for the UK, successful sequestration is expected for arable land over most of England and Wales, with unsuccessful sequestration expected in parts of Scotland, due to already carbon rich soils (existing woodland). Also, for Scotland, the relatively lower yields in this colder climate makes CO_2 negativity harder to achieve. Soils already rich in carbon includes peatland and mature forest. Grassland can also be carbon rich, however Milner et al. further argues that the most successful carbon sequestration in the UK takes place below improved grasslands. The bottom graphic displays the estimated yield necessary to achieve CO_2 negativity for different levels of existing soil carbon saturation.

The perennial rather than annual nature of Miscanthus crops implies that the significant below-ground carbon accumulation each year is allowed to continue undisturbed. No annual plowing or digging means no increased carbon oxidation and no stimulation of the microbe populations in the soil, and therefore no accelerated carbon-to-CO_2 conversion happening in the soil every spring.

Savings Comparison

Fundamentally, the below-ground carbon accumulation works as a GHG mitigation tool because it removes CO_2 from the above-ground CO_2 circulation (the circulation from plant to atmosphere and back into plant). The above-ground circulation is driven by photosynthesis and combustion – first, the miscanthus fields absorb CO_2 and assimilates it as carbon in its tissue both above and below ground. When the above-ground carbon is harvested and burned, it is released back into the atmosphere, in the form of CO_2. However, an equivalent amount of CO_2 (and possibly more, if the biomass is expanding) is absorbed back by next season's growth, and the cycle repeats. This above-ground cycle has the potential to be carbon neutral, but of course the human involvement in the process means additional energy input, often coming from fossil sources. If the fossil energy spent on the operation is high compared to the operation's energy output, the CO_2 footprint can approach, match or even exceed the CO_2 footprint from burning fossil fuels directly, as has been shown to be the case for several first-generation biofuel projects. Transport fuels might be worse than solid fuels in this regard.

The problem can be dealt with both from the perspective of increasing the amount of carbon that is moved below ground, and from the perspective of decreasing fossil fuel input to the above-ground operation. If enough carbon is moved below ground, it can compensate for a bio-energy project's total CO_2 emittance. On the other hand, if the above-ground CO_2 cost decreases, less below-ground carbon allocation is needed for the bioenergy project to become CO_2 neutral or negative.

For first generation bio-energy crops, the greenhouse gas footprints were often large, but second generation bio-energy crops like Miscanthus reduces its CO_2 footprint drastically. Hastings et al. found that Miscanthus crops "almost always has a smaller environmental footprint than first generation annual bioenergy ones." A large meta-study of 138 individual studies, done by Harris et al., revealed that second generation perennial grasses (miscanthus and switchgrass) on average stores 5 times more carbon in the ground than short rotation coppice forestry plantations (poplar and willow). Compared to fossil fuels, the GHG savings are large – without adding in the carbon sequestration negative numbers, Miscanthus fuel has a GHG cost of 0.4 - 1.6 grams CO_2-equivalents per megajoule, compared to 33 grams for coal, 22 for liquefied natural gas, 16 for North Sea gas, and 4 for wood chips imported to Britain from the USA.

Confirming the above numbers, McCalmont et al. found that the mean energy input/output ratios for Miscanthus is 10 times better than for annual crops, while GHG costs are 20-30 times better than for fossil fuels. For instance, Miscanthus chips for heating saved 22.3 tonnes of CO_2 emittance per hectare per year in the UK (9 tonnes per acre), while maize for heating and power saved 6.3 (2.5 per acre). Rapeseed for biodiesel saved only 3.2 (1.3 per acre). Lewandowski et al. found that each hectare (2.47 acres) of Central European arable land planted with Miscanthus can reduce the atmospheric CO_2 level with up to 30.6 tonnes per year, save 429 GJ of fossil energy used each year, with 78 euros earned per tonne reduced CO_2 (2387 euros earned per hectare per year) – given that the biomass is produced and used locally (within 500 km / 310 miles). For Miscanthus planted on marginal land limited by cold temperatures (Moscow), the reduction in atmospheric CO_2 is estimated to be 19.2 tonnes per hectare per year (7.7 tonnes per acre), with fossil energy savings of 273 GJ per hectare per year (110 GJ per acre). For marginal land limited by drought (Turkey), the atmospheric CO_2 level can potentially be reduced with 24 tonnes per hectare per year (9.7 tonnes per acre), with fossil energy savings of 338 GJ per hectare per year (137 tonnes per acre). Based on similar numbers, Poeplau and Don expect Miscanthus plantations to grow large in Europe in the coming decades. Whitaker et al. states that after some discussion, there is now consensus in the scientific community that "the GHG balance of perennial bioenergy crop cultivation will often be favourable" also when considering the implicit direct and indirect land use changes.

Biodiversity

Felten and Emmerling found that the number of earthworm species per square meter was 5.1 for Miscanthus, 3 for maize, and 6,4 for fallow (totally unattended land), and states that "it was clearly found that land-use intensity was the dominant regressor for earthworm abundance and total number of species." Because the extensive leaf litter on the ground helps the soil to stay moist, and also protect from predators, they conclude that "Miscanthus had quite positive effects on earthworm communities" and recommend that "Miscanthus may facilitate a diverse earthworm community even in intensive agricultural landscapes."

Nsanganwimana et al. found that the bacterial activity of certain bacteria belonging to the proteobacteria group almost doubles in the presence of M. x giganteus root exudates.

Lewandowski et al. found that young Miscanthus stands sustain high plant species diversity, but as the Miscanthus stands mature, the canopy closes, and less sunlight reach the ground. In this situation it gets harder for the weeds to survive. Lewandowski et al. found 16 different weed species per 25 m2 plot. The dense canopy works as protection for other life-forms though; Lewandowski et al. notes that "Miscanthus stands are usually reported to support farm biodiversity, providing habitat for birds, insects, and small mammals." Both Haughton et al. and Bellamy et al. found that the Miscanthus overwinter vegetative structure provided an important cover and habitat resource, with high levels of diversity in comparison with competing energy grasses. This effect was particularly evident for beetles, flies, and birds, with breeding skylarks and lapwings being recorded in the crop itself. The Miscanthus crop offers a different ecological niche for each season – the authors attribute this to the continually evolving structural heterogeneity of a Miscanthus crop, with different species finding shelter at different times during its development – woodland birds found shelter in the winter and farmland birds in the summer. For birds, 0.92 breeding pairs species per hectare (0.37 per acre) was found in the Miscanthus field, compared to 0.28 (0.11) in the wheat field. The authors note that due to the high carbon to nitrogen ratio it is in the field's margins and interspersed woodlands that the majority of the food resoures are to be found. Miscanthus fields work as barriers against chemical leaching into these key habitats however.

Water Quality

McCalmont et al. claims Miscanthus fields leads to significantly improved water quality because of significantly less nitrate leaching. Likewise, Whitaker et al. claims that there is drastically reduced nitrate leaching from Miscanthus fields compared to the typical maize/soy rotation because of low or zero fertilizer requirements, the continuous presence of a plant root sink for nitrogen, and the efficient internal recycling of nutrients by perennial grass species. For instance, a recent meta-study concluded that Miscanthus had nine times less subsurface loss of nitrate compared to maize or maize grown in rotation with soya bean.

Soil Quality

The fibrous, extensive Miscanthus rooting system and the lack of tillage disturbance improves infiltration, hydraulic conductivity and water storage compared to annual row crops, and results in the porous and low bulk density soil typical under perennial grasses, with water holding capabilities expected to increase by 100-150 mm. Nsanganwimana et al. argues that Miscanthus improves carbon input to the soil, and promote microorganism activity and diversity, which are important for soil particle aggregation and rehabilitation processes. On a former fly ash deposit site, with alkaline pH, nutrient deficiency, and little water-holding capacity, a Miscanthus x giganteus crop was successfully established – in the sense that the roots and rhizomes grew quite well, supporting and enhancing nitrification processes, although the above-ground dry weight yield was low because of the conditions. The authors argue that Miscanthus' ability to improve soil quality, even on contaminated land, can be used in combination with organic amendments on soils with a low agronomic value. For instance, there is a great potential to increase yield on contaminated marginal land low in nutrients by fertilizing it with nutrient-rich sewage sludge or wastewater. The authors claim that this practice offer the three-fold advantage of improving soil productivity, increasing biomass yields, and reducing costs for treatment and disposal of sewage sludge in line with the specific legislation in each country.

Invasiveness

Miscanthus x giganteus' parents on both sides, M. sinensis and M. sacchariflorus, are both potentially invasive species, because they both produce viable seeds. M. x giganteus does not produce viable seeds however, and Nsanganwimana et al. claims that "there has been no report on the threat of invasion due to rhizome growth extension from long-term commercial plantations to neighboring arable land."

References

- Definition-of-feedstock: thoughtco.com, Retrieved 15 May, 2019

- Ara Kirakosyan; Peter B. Kaufman (2009-08-15). Recent Advances in Plant Biotechnology. P. 169. ISBN 9781441901934. Retrieved 14 February 2013

- Camelina-for-biofuel-production: extension.org, Retrieved 19 July, 2019

- Janick, Jules; Robert E. Paull (2008). The Encyclopedia of Fruit & Nuts. CABI. Pp. 371–372. ISBN 978-0-85199-638-7

- "Jatropha curcas". Germplasm Resources Information Network(GRIN). Agricultural Research Service (ARS), United States Department of Agriculture (USDA). Retrieved 2010-10-14

- Carroll, Andrew; Somerville, Chris (June 2009). "Cellulosic Biofuels". Annual Review of Plant Biology. 60 (1): 165–182. Doi:10.1146/annurev.arplant.043008.092125

- Switchgrass-panicum-virgatum-for-biofuel-production: extension.org, Retrieved 15 June, 2019

- Barbara A. Tokay "Biomass Chemicals" in Ullmann's Encyclopedia of Industrial Chemistry 2002, Wiley-VCH, Weinheim. Doi:10.1002/14356007.a04-099

Production Processes and Technologies

5

- **Dark Fermentation**
- **Photofermentation**
- **Ethanol Fermentation**
- **Biodiesel Production**
- **Vegetable Oil Refinin**
- **Energy Forestry**
- **Gasificatio**
- **Methanol Economy**
- **Ethanol Fuel Energy Balance**

Some of the most common processes and techniques for biofuel production include dark fermentation, photofermentation, vegetable oil refining and gasification. The chapter closely examines these key concepts, technologies and processes related to biofuel production to provide an extensive understanding of the subject.

Dark Fermentation

Dark fermentation is the fermentative conversion of organic substrate to biohydrogen. It is a complex process manifested by diverse groups of bacteria, involving a series of biochemical reactions using three steps similar to anaerobic conversion. Dark fermentation differs from photofermentation in that it proceeds without the presence of light.

Fermentative/hydrolytic microorganisms hydrolyze complex organic polymers to monomers which are further converted to a mixture of lower-molecular-weight organic acids and alcohols by obligatory producing acidogenic bacteria.

Utilization of wastewater as a potential substrate for biohydrogen production has been drawing considerable interest in recent years especially in the dark fermentation process. Industrial wastewater as a fermentative substrate for H_2 production addresses most of the criteria required for substrate selection viz., availability, cost and biodegradability. Chemical wastewater, cattle wastewater, dairy process wastewater, starch hydrolysate wastewater and designed synthetic wastewater have been reported to produce biohydrogen apart from wastewater treatment from dark fermentation processes using selectively enriched mixed cultures under acidophilic conditions. Various wastewaters viz., paper mill wastewater, starch effluent, food processing wastewater, domestic wastewater, rice winery wastewater, distillery and molasses based wastewater, wheat straw wastes and palm oil mill wastewater have been studied as fermentable substrates for H_2 production along with wastewater treatment. Using wastewater as a fermentable substrate facilitates both wastewater treatment apart from H_2 production. The efficiency of the dark fermentative H_2 production process was found to depend on pre-treatment of the mixed consortia used as a biocatalyst, operating pH, and organic loading rate apart from wastewater characteristics.

In spite of its advantages, the main challenge observed with fermentative H_2 production processes is the relatively low energy conversion efficiency from the organic source. Typical H_2 yields range from 1 to 2 mol of H_2/mol of glucose, which results in 80-90% of the initial COD remaining in the wastewater in the form of various volatile organic acids (VFAs) and solvents, such as acetic acid, propionic acid, butyric acid, and ethanol. Even under optimal conditions about 60-70% of the original organic matter remains in solution. Bioaugmentation with selectively enriched acidogenic consortia to enhance H_2 production was also reported. Generation and accumulation of soluble acid metabolites causes a sharp drop in the system pH and inhibits the H_2 production process. Usage of unutilized carbon sources present in acidogenic process for additional biogas production sustains the practical applicability of the process. One way to utilize/recover the remaining organic matter in a usable form is to produce additional H_2 by terminal integration of photo-fermentative processes of H_2 production and methane by integrating acidogenic processes to terminal methanogenic processes.

Photofermentation

Photofermentation is the fermentative conversion of organic substrate to biohydrogen manifested by a diverse group of photosynthetic bacteria by a series of biochemical reactions involving three steps similar to anaerobic conversion. Photofermentation differs from dark fermentation because it only proceeds in the presence of light.

For example, photo-fermentation with *Rhodobacter sphaeroides* SH2C (or many other purple non-sulfur bacteria) can be employed to convert small molecular fatty acids into hydrogen and other products.

Light-dependent Pathways

Phototropic Bacteria

Phototropic bacteria produce hydrogen gas via photofermentation, where the hydrogen is sourced from organic compounds.

$$C_6H_{12}O_6 + 6H_2O \xrightarrow{h\upsilon} 6CO_2 + 12H_2$$

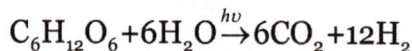

Photolytic Producers

Photolytic producers are similar to phototrophs, but source hydrogen from water molecules that are broken down as the organism interacts with light. Photolytic producers consist of algae and certain photosynthetic bacteria.

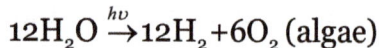

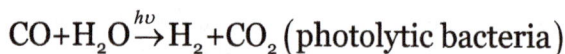

$$12H_2O \xrightarrow{h\upsilon} 12H_2 + 6O_2 \text{ (algae)}$$

$$CO + H_2O \xrightarrow{h\upsilon} H_2 + CO_2 \text{ (photolytic bacteria)}$$

Sustainable Energy Production

Photofermentation via purple nonsulfur producing bacteria has been explored as a method for the production of biofuel. The natural fermentation product of these bacteria, hydrogen gas, can be harnessed as a natural gas energy source. Photofermentation via algae instead of bacteria is used for bioethanol production, among other liquid fuel alternatives.

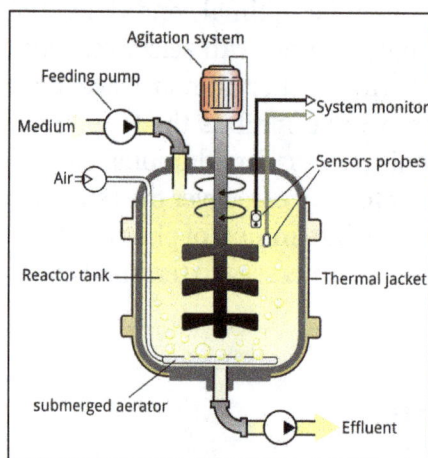

Basic principles of a bioreactor. The photofermentation bioreactor would not include an air pathway.

Mechanism

The bacteria and their energy source are held in a bioreactor chamber that is impermeable to air and oxygen free. The proper temperature for the bacterial species is maintained in the bioreactor. The bacteria are sustained with a carbohydrate diet consisting of simple saccharide molecules. The carbohydrates are typically sourced from agricultural or forestry waste.

Variations

In addition to wild type forms of *Rhodopseudomonas palustris*, scientists have used genetically modified forms to produce hydrogen as well. Other explorations include expanding the bioreactor system to hold a combination of bacteria, algae or cyanobacteria. Ethanol production is performed by the algae *Chlamydomonas reinhardtii*, among other species, in cycling light and dark environments. The cycling of light and dark environments has also been explored with bacteria for hydrogen production, increasing hydrogen yield.

Depiction of algae (species not specified) in a bioreactor suitable for bioethanol production.

Advantages

The bacteria are typically fed with broken down agricultural waste or undesired crops, such as water lettuce or sugar beet molasses. The high abundance of such waste ensures the stable food source for the bacteria and productively uses human-produced waste. In comparison with dark fermentation, photofermentation produces more hydrogen per reaction and avoids the acidic end products of dark fermentation.

Limitations

The primary limitations of photofermentation as a sustainable energy source stem from the precise requirements of maintaining the bacteria in the bioreactor. Researchers have found it difficult to maintain a constant temperature for the bacteria within the bioreactor. Furthermore, the growth media for the bacteria must be rotated and refreshed without introducing air to the bioreactor system, complicating the already expensive bioreactor set up.

Ethanol Fermentation

Ethanol fermentation, also called alcoholic fermentation, is a biological process which converts sugars such as glucose, fructose, and sucrose into cellular energy, producing ethanol and carbon dioxide as by-products.This is because yeasts perform this conversion in the absence of oxygen, alcoholic fermentation is considered an anaerobic process. It also takes place in some species of fish (including goldfish and carp) where (along with lactic acid fermentation) it provides energy when oxygen is scarce.

Ethanol fermentation has many uses, including the production of alcoholic beverages, the production of ethanol fuel, and bread cooking.

Biochemical Process of Fermentation of Sucrose

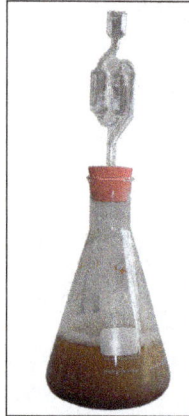

A laboratory vessel being used for the fermentation of straw.

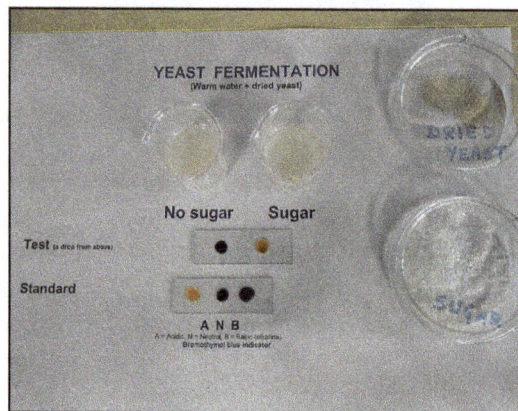

Fermentation of sucrose by yeast.

The chemical equations below summarize the fermentation of sucrose ($C_{12}H_{22}O_{11}$) into ethanol (C_2H_5OH). Alcoholic fermentation converts one mole of glucose into two moles of ethanol and two moles of carbon dioxide, producing two moles of ATP in the process.

The overall chemical formula for alcoholic fermentation is:

$C_6H_{12}O_6 \rightarrow 2\ C_2H_5OH + 2\ CO_2$

Sucrose is a dimer of glucose and fructose molecules. In the first step of alcoholic fermentation, the enzyme invertase cleaves the glycosidic linkage between the glucose and fructose molecules.

$C_{12}H_{22}O_{11} + H_2O + invertase \rightarrow 2\ C_6H_{12}O_6$

Next, each glucose molecule is broken down into two pyruvate molecules in a process known as glycolysis. Glycolysis is summarized by the equation:

$C_6H_{12}O_6 + 2\ ADP + 2\ P_i + 2\ NAD^+ \rightarrow 2\ CH_3COCOO^- + 2\ ATP + 2\ NADH + 2\ H_2O + 2\ H^+$

CH_3COCOO^- is pyruvate, and P_i is inorganic phosphate. Finally, pyruvate is converted to ethanol and CO_2 in two steps, regenerating oxidized NAD+ needed for glycolysis:

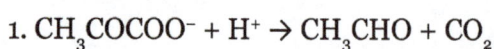

1. $CH_3COCOO^- + H^+ \rightarrow CH_3CHO + CO_2$

catalyzed by pyruvate decarboxylase

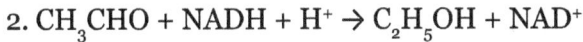

2. $CH_3CHO + NADH + H^+ \rightarrow C_2H_5OH + NAD^+$

This reaction is catalyzed by alcohol dehydrogenase (ADH1 in baker's yeast).

As shown by the reaction equation, glycolysis causes the reduction of two molecules of NAD^+ to NADH. Two ADP molecules are also converted to two ATP and two water molecules via substrate-level phosphorylation.

Related Processes

Fermentation of sugar to ethanol and CO_2 can also be done by *Zymomonas mobilis*, however the path is slightly different since formation of pyruvate does not happen by glycolysis but instead by the Entner–Doudoroff pathway. Other microorganisms can produce ethanol from sugars by fermentation but often only as a side product. Examples are:

- Heterolactic acid fermentation in which Leuconostoc bacterias produce Lactate + Ethanol + CO_2.

- Mixed acid fermentation where Escherichia produce ethanol mixed with lactate, acetate, succinate, formate, CO_2, and H_2.

- 2,3-butanediol fermentation by Enterobacter producing ethanol, butanediol, lactate, formate, CO_2, and H_2.

Effect of Oxygen

Fermentation does not require oxygen. If oxygen is present, some species of yeast (e.g., *Kluyveromyces lactis* or *Kluyveromyces lipolytica*) will oxidize pyruvate completely to carbon dioxide and water in a process called cellular respiration, hence these species of yeast will produce ethanol only in an anaerobic environment (not cellular respiration). This phenomenon is known as the Pasteur effect.

However, many yeasts such as the commonly used baker's yeast *Saccharomyces cerevisiae* or fission yeast *Schizosaccharomyces pombe* under certain conditions, ferment rather than respire even in the presence of oxygen. In wine making this is known as the counter-Pasteur effect. These yeasts will produce ethanol even under aerobic conditions, if they are provided with the right kind of nutrition. During batch fermentation, the rate of ethanol production per milligram of cell protein is maximal for a brief period early in this process and declines progressively as ethanol accumulates in the surrounding broth. Studies demonstrate that the removal of this accumulated ethanol does not immediately restore fermentative activity, and they provide evidence that the decline in metabolic rate is due to physiological changes (including possible ethanol damage) rather than to the presence of ethanol. Several potential causes for the decline in fermentative activity have been investigated. Viability remained at or above 90%, internal pH remained near neutrality, and the specific activities of the glycolytic and alcohologenic enzymes (measured in vitro) remained high throughout batch fermentation. None of these factors appears to be causally related to the fall in fermentative activity during batch fermentation.

Bread Baking

Ethanol fermentation causes bread dough to rise. Yeast organisms consume sugars in the dough and produce ethanol and carbon dioxide as waste products. The carbon dioxide forms bubbles in the dough, expanding it to a foam. Less than 2% ethanol remains after baking.

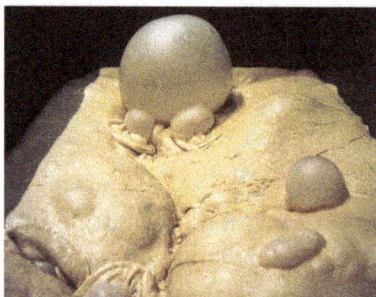

The formation of carbon dioxide — a byproduct of ethanol fermentation — causes bread to rise.

Alcoholic Beverages

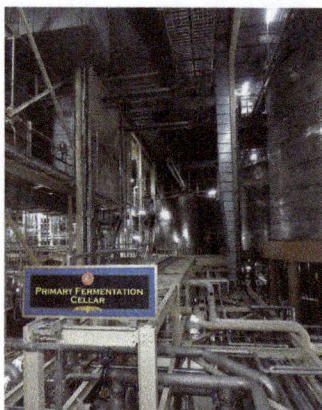

Primary fermentation cellar.

All ethanol contained in alcoholic beverages (including ethanol produced by carbonic maceration) is produced by means of fermentation induced by yeast.

- Wine is produced by fermentation of the natural sugars present in grapes; cider and perry are produced by similar fermentation of natural sugar in apples and pears, respectively; and other fruit wines are produced from the fermentation of the sugars in any other kinds of fruit. Brandy and eaux de vie (e.g. slivovitz) are produced by distillation of these fruit-fermented beverages.

- Mead is produced by fermentation of the natural sugars present in honey.

- Beer, whiskey, and vodka are produced by fermentation of grain starches that have been converted to sugar by the enzyme amylase, which is present in grain kernels that have been malted (i.e. germinated). Other sources of starch (e.g. potatoes and unmalted grain) may be added to the mixture, as the amylase will act on those starches as well. Whiskey and vodka are also distilled; gin and related beverages are produced by the addition of flavoring agents to a vodka-like feedstock during distillation.

- Rice wines (including sake) are produced by the fermentation of grain starches converted to sugar by the mold *Aspergillus oryzae. Baijiu, soju,* and *shōchū* are distilled from the product of such fermentation.

- Rum and some other beverages are produced by fermentation and distillation of sugarcane. Rum is usually produced from the sugarcane product molasses.

In all cases, fermentation must take place in a vessel that allows carbon dioxide to escape but prevents outside air from coming in. This is to reduce risk of contamination of the brew by unwanted bacteria or mold and because a buildup of carbon dioxide creates a risk the vessel will rupture or fail, possibly causing injury or property damage.

Feedstocks for Fuel Production

Yeast fermentation of various carbohydrate products is also used to produce the ethanol that is added to gasoline.

The dominant ethanol feedstock in warmer regions is sugarcane. In temperate regions, corn or sugar beets are used.

In the United States, the main feedstock for the production of ethanol is currently corn. Approximately 2.8 gallons of ethanol are produced from one bushel of corn (0.42 liter per kilogram). While much of the corn turns into ethanol, some of the corn also yields by-products such as DDGS (distillers dried grains with solubles) that can be used as feed for livestock. A bushel of corn produces about 18 pounds of DDGS (320 kilograms of DDGS per metric ton of maize). Although most of the fermentation plants have been built in corn-producing regions, sorghum is also an important feedstock for ethanol production in the Plains states. Pearl millet is showing promise as an ethanol feedstock for the southeastern U.S. and the potential of duckweed is being studied.

In some parts of Europe, particularly France and Italy, grapes have become a *de facto* feedstock for fuel ethanol by the distillation of surplus wine. Surplus sugary drinks may also be used. In Japan, it has been proposed to use rice normally made into sake as an ethanol source.

Cassava as Ethanol Feedstock

Ethanol can be made from mineral oil or from sugars or starches. Starches are cheapest. The starchy crop with highest energy content per acre is cassava, which grows in tropical countries.

Thailand already had a large cassava industry in the 1990s, for use as cattle feed and as a cheap admixture to wheat flour. Nigeria and Ghana are already establishing cassava-to-ethanol plants. Production of ethanol from cassava is currently economically feasible when crude oil prices are above US$120 per barrel.

New varieties of cassava are being developed, so the future situation remains uncertain. Currently, cassava can yield between 25-40 tonnes per hectare (with irrigation and fertilizer), and from a tonne of cassava roots, circa 200 liters of ethanol can be produced (assuming cassava with 22% starch content). A liter of ethanol contains circa 21.46 MJ of energy. The overall energy efficiency of cassava-root to ethanol conversion is circa 32%.

The yeast used for processing cassava is *Endomycopsis fibuligera*, sometimes used together with bacterium *Zymomonas mobilis*.

Byproducts of Fermentation

Ethanol fermentation produces unharvested byproducts such as heat, carbon dioxide, food for livestock, water, methanol, fuels, fertilizer and alcohols. The cereal unfermented solid residues from the fermentation process, which can be used as livestock feed or in the production of biogas, are referred to as Distillers grains and sold as WDG, *Wet Distiller's grains*, and DDGS, *Dried Distiller's Grains with Solubles*, respectively.

Microbes used in ethanol fermentation:

- Yeast

    ◦ Saccharomyces cerevisiae

    ◦ Schizosaccharomyces

- Zymomonas mobilis (a bacterium)

Biodiesel Production

Biodiesel production is the process of producing the biofuel, biodiesel, through the chemical reactions of transesterification and esterification. This involves vegetable or animal fats and oils being reacted with short-chain alcohols (typically methanol or ethanol). The alcohols used should be of low molecular weight. Ethanol is the most used because of its low cost, however, greater conversions into biodiesel can be reached using methanol. Although the transesterification reaction can be catalyzed by either acids or bases, the base-catalyzed reaction is more common. This path has lower reaction times and catalyst cost than those acid catalysis. However, alkaline catalysis has the disadvantage of high sensitivity to both water and free fatty acids present in the oils.

Process Steps

The major steps required to synthesize biodiesel are as follows:

Feedstock Pretreatment

Common feedstock used in biodiesel production include yellow grease (recycled vegetable oil), "virgin" vegetable oil, and tallow. Recycled oil is processed to remove impurities from cooking, storage, and handling, such as dirt, charred food, and water. Virgin oils are refined, but not to a food-grade level. Degumming to remove phospholipids and other plant matter is common, though refinement processes vary. Water is removed as its presence during base-catalyzed transesterification will cause the triglycerides to hydrolyze, producing soap instead of biodiesel.

A sample of the cleaned feedstock is then tested, via titration against a standardized base solution, in order to determine the concentration of free fatty acids present in the vegetable oil sample. These acids are then either removed, typically through neutralization, or esterified, into biodiesel or glycerides.

Reactions

Base-catalyzed transesterification reacts lipids (fats and oils) with alcohol (typically methanol or ethanol) to produce biodiesel and an impure coproduct, glycerol. If the feedstock oil is used or has a high acid content, acid-catalyzed esterification can be used to react fatty acids with alcohol to produce biodiesel. Other methods, such as fixed-bed reactors, supercritical reactors, and ultrasonic reactors, forgo or decrease the use of chemical catalysts.

Product Purificatio

Products of the reaction include not only biodiesel, but also byproducts, soap, glycerol, excess alcohol, and trace amounts of water. All of these byproducts must be removed to meet the standards, but the order of removal is process-dependent.

The density of glycerol is greater than that of biodiesel, and this property difference is exploited to separate the bulk of the glycerol coproduct. Residual methanol is typically recovered by distillation and reused. Soaps can be removed or converted into acids. Residual water is also removed from the fuel.

Reactions

Transesterificatio

Animal and plant fats and oils are composed of triglycerides, which are esters formed by the reactions of three free fatty acids and the trihydric alcohol, glycerol. In the transesterification process, the added alcohol (commonly, methanol or ethanol) is deprotonated with a base to make it a stronger nucleophile. As can be seen, the reaction has no other inputs than the triglyceride and the alcohol. Under normal conditions, this reaction will proceed either exceedingly slowly or not at all, so heat, as well as catalysts (acid and/or base) are used to speed the reaction. It is important to note that the acid or base are not consumed by the transesterification reaction, thus they are not reactants, but catalysts. Common catalysts for transesterification include sodium hydroxide, potassium hydroxide, and sodium methoxide.

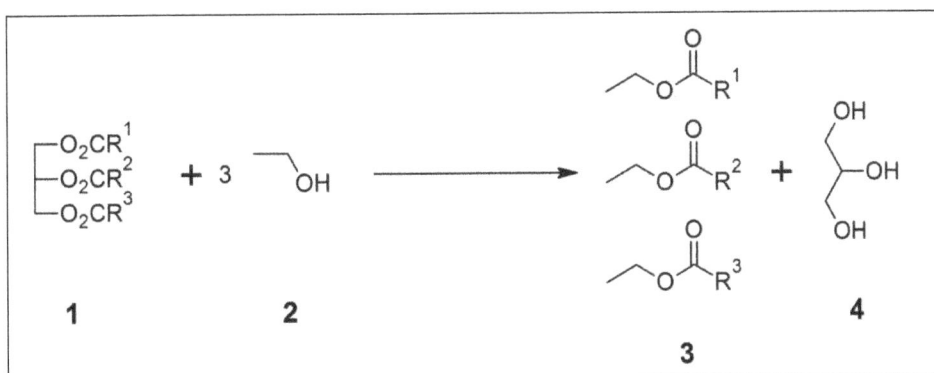

Triglycerides (1) are reacted with an alcohol such as ethanol (2) to give ethyl esters of fatty acids (3) and glycerol (4).

Almost all biodiesel is produced from virgin vegetable oils using the base-catalyzed technique as it is the most economical process for treating virgin vegetable oils, requiring only low temperatures and pressures and producing over 98% conversion yield (provided the starting oil is low in moisture and free fatty acids). However, biodiesel produced from other sources or by other methods may require acid catalysis, which is much slower.

The alcohol reacts with the fatty acids to form the mono-alkyl ester (biodiesel) and crude glycerol. The reaction between the biolipid (fat or oil) and the alcohol is a reversible reaction so excess alcohol must be added to ensure complete conversion.

Base-catalysed Transesterificatio Mechanism

The transesterification reaction is base catalyzed. Any strong base capable of deprotonating the alcohol will do (e.g. NaOH, KOH, sodium methoxide, etc.), but the sodium and potassium hydroxides are often chosen for their cost. The presence of water causes undesirable base hydrolysis, so the reaction must be kept dry.

In the transesterification mechanism, the carbonyl carbon of the starting ester (RCOOR^1) undergoes nucleophilic attack by the incoming alkoxide (R^{2}O$^-$) to give a tetrahedral intermediate, which either reverts to the starting material, or proceeds to the transesterified product (RCOOR^2). The various species exist in equilibrium, and the product distribution depends on the relative energies of the reactant and product.

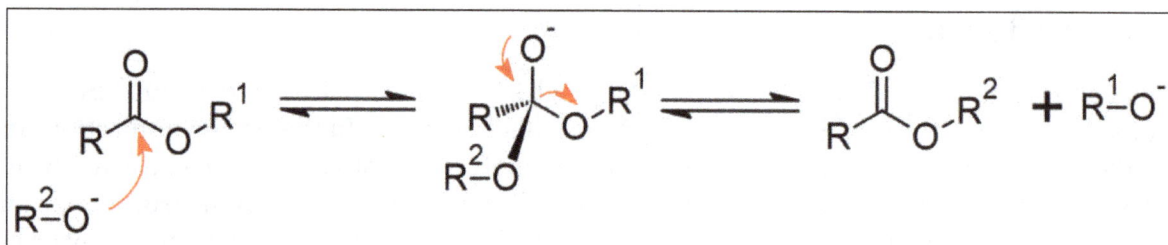

Production Methods

Supercritical Process

An alternative, catalyst-free method for transesterification uses supercritical methanol at high temperatures and pressures in a continuous process. In the supercritical state, the oil and methanol are in a single phase, and reaction occurs spontaneously and rapidly. The process can tolerate water in the feedstock, free fatty acids are converted to methyl esters instead of soap, so a wide variety of feedstocks can be used. Also the catalyst removal step is eliminated. High temperatures and pressures are required, but energy costs of production are similar or less than catalytic production routes.

Ultra- and High Shear in-line and Batch Reactors

Ultra- and High Shear in-line or batch reactors allow production of biodiesel continuously, semi-continuously, and in batch-mode. This drastically reduces production time and increases production volume.

The reaction takes place in the high-energetic shear zone of the Ultra- and High Shear mixer by reducing the droplet size of the immiscible liquids such as oil or fats and methanol. Therefore, the smaller the droplet size the larger the surface area the faster the catalyst can react.

Ultrasonic Reactor Method

In the ultrasonic reactor method, the ultrasonic waves cause the reaction mixture to produce and collapse bubbles constantly. This cavitation simultaneously provides the mixing and heating required to carry out the transesterification process. Thus, using an ultrasonic reactor for biodiesel production drastically reduces the reaction time, reaction temperatures, and energy input. Hence the process of transesterification can run inline rather than using the time consuming batch processing. Industrial scale ultrasonic devices allow for the industrial scale processing of several thousand barrels per day.

Lipase-catalyzed Method

Large amounts of research have focused recently on the use of enzymes as a catalyst for the transesterification. Researchers have found that very good yields could be obtained from crude and used oils using lipases. The use of lipases makes the reaction less sensitive to high free fatty-acid content, which is a problem with the standard biodiesel process. One problem with the lipase reaction is that methanol cannot be used because it inactivates the lipase catalyst after one batch. However, if methyl acetate is used instead of methanol, the lipase is not in-activated and can be used for several batches, making the lipase system much more cost effective.

Volatile Fatty Acids from Anaerobic Digestion of Waste Streams

Lipids have been drawing considerable attention as a substrate for biodiesel production owing to its sustainability, non-toxicity and energy efficient properties. However, due to cost reasons, attention must be focused on the non-edible sources of lipids, in particular oleaginous microorganisms. Such microbes have the ability to assimilate the carbon sources from a medium and convert the carbon into lipid storage materials. The lipids accumulated by these oleaginous cells can then be transesterified to form biodiesel.

Vegetable Oil Refinin

Vegetable oil refining is a process to transform vegetable oil into biofuel by hydrocracking or hydrogenation. Hydrocracking breaks big molecules into smaller ones using hydrogen while hydrogenation adds hydrogen to molecules. These methods can be used for production of gasoline, diesel, propane, and other chemical feedstock. Diesel fuel produced from these sources is known as *green diesel* or *renewable diesel*.

Feedstock

The majority of plant and animal oils are vegetable oils which are triglycerides—suitable for refining. Refinery feedstock includes canola, algae, jatropha, salicornia, palm oil, and tallow. One type of algae, Botryococcus braunii produces a different type of oil, known as a triterpene, which is transformed into alkanes by a different process.

Comparison to Biodiesel

Based on its feedstock green diesel could be classified as biodiesel; however, based on the processing technology and chemical formula green diesel and biodiesel are different products. The chemical reaction commonly used to produce biodiesel is known as transesterification. Vegetable oil and alcohol are reacted, producing esters, or biodiesel, and the coproduct, glycerol.

When refining vegetable oil, no glycerol is produced, only fuels.

Commercialization

Various stages of converting renewable hydrocarbon fuels produced by hydrotreating is done throughout energy industry. Some commercial examples of vegetable oil refining are NExBTL, H-Bio, the ConocoPhilips process, and the UOP/Eni Ecofining process. Neste Oil is the largest manufacturer, producing 2 million tons annually (2013). Neste Oil completed their first NExBTL plant in the summer 2007 and the second one in 2009. Petrobras planned to use 256 megalitres (1,610,000 bbl) of vegetable oils in the production of H-Bio fuel in 2007. ConocoPhilips is processing 42,000 US gallons per day (1,000 bbl/d) of vegetable oil.Other companies working on the commercialization and industrialization of renewable hydrocarbons and biofuels include Neste, REG Synthetic Fuels, LLC, ENI, UPM Biofuels, Diamond Green Diesel partnered with countries across the globe.In practice, these renewable diesels lower greenhouse gas emissions by 40-90%, have higher energy per content yields than petroleum-based diesels, and better cold-flow properties to work in colder climates. In addition, all of these green diesels can be introduced into any diesel engine or infrastructure without many mechanical modifications at any ratio with petroleum-based diesels.

Renewable diesel from vegetable oil in particular is a growing and profound substitute for petroleum. California fleets used over 200,000,000 gallons of renewable diesel in 2017. CARB predicts over 2 billion gallons of fuel to be consumed in the state under its Low Carbon Fuel Standard requirements in the next ten years. Fleets operating on Renewable Diesel from various refiners and feedstocks are reported to see lower emissions, reduced maintenance costs, and nearly identical experience when driving with this fuel.

Energy Forestry

Energy forestry is a form of forestry in which a fast-growing species of tree or woody shrub is grown specifically to provide biomass or biofuel for heating or power generation.

The two forms of energy forestry are short rotation coppice and short rotation forestry:

- Short rotation coppice may include tree crops of poplar, willow or eucalyptus, grown for two to five years before harvest.

- Short rotation forestry are crops of alder, ash, birch, eucalyptus, poplar, and sycamore, grown for eight to 20 years before harvest.

Benefit

The main advantage of using "grown fuels", as opposed to fossil fuels such as coal, natural gas and oil, is that while they are growing they absorb the near-equivalent in carbon dioxide (an important greenhouse gas) to that which is later released in their burning. In comparison, burning fossil fuels increases atmospheric carbon unsustainably, by using carbon that was added to the Earth's carbon sink millions of years ago. This is a prime contributor to climate change.

According to the FAO, compared to other energy crops, wood is among the most efficient sources of bioenergy in terms of quantity of energy released by unit of carbon emitted. Other advantages of generating energy from trees, as opposed to agricultural crops, are that trees do not have to be harvested each year, the harvest can be delayed when market prices are down, and the products can fulfil a variety of end-uses.

Yields of some varieties can be as high as 12 oven dry tonnes every year. However, commercial experience on plantations in Scandinavia have shown lower yield rates.

These crops can also be used in bank stabilisation and phytoremediation. In fact, experiments in Sweden with willow plantations have proved to have many beneficial effects on the soil and water quality when compared to conventional agricultural crops (such as cereal).

Problems

Although in many areas of the world government funding is still required to support large scale development of energy forestry as an industry, it is seen as a valuable component of the renewable energy network and will be increasingly important in the future.

The system of energy forestry has faced criticism over food vs. fuel, whereby it has become financially profitable to replace food crops with energy crops. It has to be noted, however, that such energy forests do not necessarily compete with food crops for highly productive land as they can be grown on slopes, marginal, or degraded land as well - sometimes even with long-term restoration purposes in mind.

Biorefiner

A biorefinery is a facility that integrates biomass conversion processes and equipment to produce fuels, power, and value-added chemicals from biomass. Biorefinery is analogous to today's petroleum refinery, which produces multiple fuels and products from petroleum. By producing several products, a biorefinery takes advantage of the various components in biomass and their intermediates, therefore maximizing the value derived from the biomass feedstock.

A biorefinery could, for example, produce one or several low-volume, but high-value, chemical products and a low-value, but high-volume liquid transportation fuel such as biodiesel or bioethanol. At the same time, it can generate electricity and process heat, through CHP technology, for its own use and perhaps enough for sale of electricity to the local utility. The high value products increase profitability, the high-volume fuel helps meet energy needs, and the power production helps to lower energy costs and reduce GHG emissions from traditional power plant facilities.

Biorefiner Platforms

There are several platforms which can be employed in a biorefinery with the major ones being the sugar platform and the thermochemical platform (also known as syngas platform).

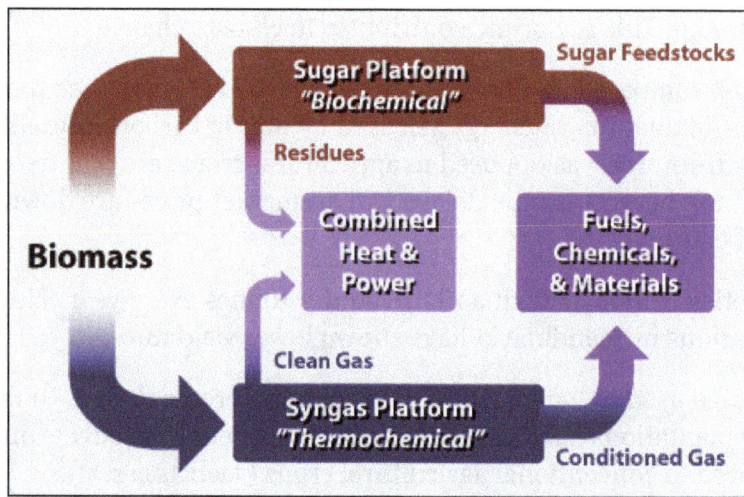

Sugar platform biorefineries breaks down biomass into different types of component sugars for fermentation or other biological processing into various fuels and chemicals. On the other hand, thermochemical biorefineries transform biomass into synthesis gas (hydrogen and carbon monoxide) or pyrolysis oil.

The thermochemical biomass conversion process is complex, and uses components, configurations, and operating conditions that are more typical of petroleum refining. Biomass is converted into syngas, and syngas is converted into an ethanol-rich mixture. However, syngas created from biomass contains contaminants such as tar and sulphur that interfere with the conversion of the syngas into products. These contaminants can be removed by tar-reforming catalysts and catalytic reforming processes. This not only cleans the syngas, it also creates more of it, improving process economics and ultimately cutting the cost of the resulting ethanol.

Advantages

Biorefineries can help in utilizing the optimum energy potential of organic wastes and may also resolve the problems of waste management and GHGs emissions. Biomass wastes can be converted, through appropriate enzymatic/chemical treatment, into either gaseous or liquid fuels. The pre-treatment processes involved in biorefining generate products like paper-pulp, HFCS, solvents, acetate, resins, laminates, adhesives, flavour chemicals, activated carbon, fuel enhancers, undigested sugars etc. which generally remain untapped in the traditional processes. The suitability of this process is further enhanced from the fact that it can utilize a variety of biomass resources, whether plant-derived or animal-derived.

Gasificatio

Gasification is a technology that converts carbon-containing materials, including coal, waste and biomass, into synthetic gas which in turn can be used to produce electricity and other valuable products, such as chemicals, fuels, and fertilizers.

Gasification does not involve combustion, but instead uses little or no oxygen or air in a closed reactor to convert carbon-based materials directly into a synthetic gas, or syngas.

Gasification can recover the energy locked in biomass and municipal solid waste, converting those materials into valuable products and eliminating the need for incineration or landfilling. Metals and glass should be segregated from the waste stream prior to being sent into the gasification process. The gasification process breaks these carbon-containing materials down to the molecular level, so impurities like nitrogen, sulphur, and mercury can be easily removed and sold as valuable industrial commodities.

While syngas is the primary product of the gasification plants, marketable products obtained from syngas include chemicals (45%), liquid fuels (28%), gaseous fuels (8%), and electric power (19%).

In some coal combustion-based power plants, only a third of the energy value of coal is actually converted into electricity. A coal gasification power plant, however, typically gets dual duty from the gases it produces. First, the coal gases, cleaned of impurities, are fired in a gas turbine - much like natural gas - to generate one source of electricity. The hot exhaust of the gas turbine, and some of the heat generated in the gasification process, are then used to generate steam for use in a steam turbine-generator. This dual source of electric power, called a "combined cycle," is much more efficient in converting coal's energy into usable electricity. The fuel efficiency of a coal gasification power plant in this type of combined cycle can potentially be boosted to 50 percent or more.

Carbon dioxide is emitted as a concentrated gas stream in syngas at high pressure. In this form, it can be captured and sequestered more easily and at lower costs. By contrast, when coal burns or is reacted in air, 79 percent of which is nitrogen, the resulting carbon dioxide is diluted and more costly to separate.

Gasificatio Working

Gasificatio as Incomplete Combustion

Gasification is most simply thought of as choked combustion or incomplete combustion. It is burning solid fuels like wood or coal without enough air to complete combustion, so the output gas still has combustion potential. The unburned gas is then piped away to burn elsewhere as needed.

Gas produced by this method goes by a variety of names: wood gas, syngas, producer gas, town gas, generator gas, and others. It's sometimes also called biogas, though biogas more typically refers to gas produced via microbes in anaerobic digestion. In the context of biomass gasification using air-aspirated gasifiers, the term producer gas is the term we will be using, since the other terms have implications that do not necessarily apply to the gas produced by our gasifiers.

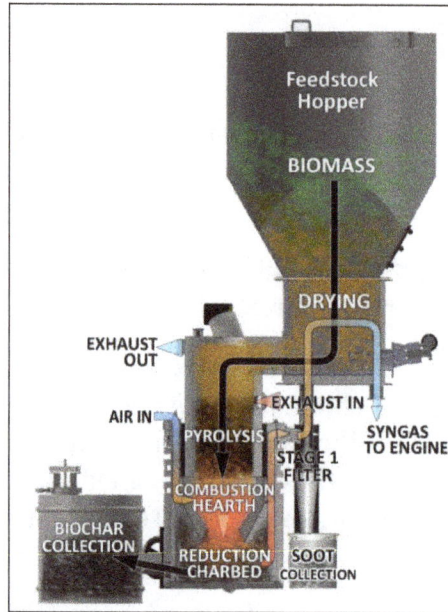

The Five Processes of Gasificatio

Gasification is more accurately understood as staged combustion. It is a series of distinct thermal events put together so as to purpose convert solid organic matter into specific hydrocarbon gases as output.

Simple incomplete combustion is a dirty mess. The goal in gasification is to take control of the discrete thermal processes usually mixed together in combustion, and reorganize them towards desired end products. In digital terms, "Gasification is the operating system of fire". Once you understand its underlying code, you can pull fire apart and reassemble it to your will, as well as a stunning variety of end products and processes.

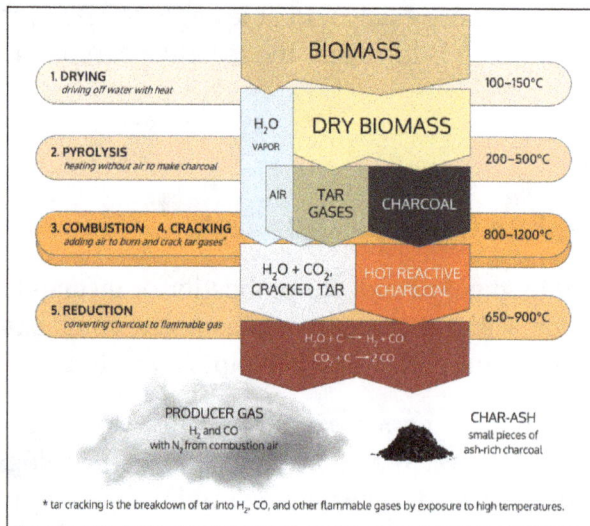

Gasification is made up for five discrete thermal processes: Drying, Pyrolysis, Combustion, Cracking, and Reduction. All of these processes are naturally present in the flame you see burning off a match, though they mix in a manner that renders them invisible to eyes not yet initiated into the mysteries of gasification. Gasification is merely the technology to pull apart and isolate these separate processes, so that we might interrupt the "fire" and pipe the resulting gases elsewhere.

Three of these processes tend to confuse all newcomers to gasification. Once you understand these three processes, all the others pieces fall in place quickly. These three non-obvious processes are Pyrolysis, Cracking, and Reduction.

Pyrolysis

Pyrolysis is the application of heat to raw biomass, in an absence of air, so as to break it down into charcoal and various tar gasses and liquids. It is essentially the process of charring.

Biomass begins to rapidly decompose with heat once its temperature rises above around 240 °C. The biomass breaks down into a combination of solids, liquids and gasses. The solids that remain we commonly call charcoal. The gasses and liquids that are released we collectively call tars.

The gasses and liquids produced during lower temp pyrolysis are simply fragments of the original biomass that break off with heat. These fragments are the more complicated H, C and O molecules in the biomass that we collectively refer to as volatiles. As the name suggests, volatiles are reactive. Or more accurately, they are less strongly bonded in the biomass than the fixed carbon, which is the direct C to C bonds.

The input to gasification is some form of solid carbonaceous material— typically biomass or coal. All organic carbonaceous material is made up of carbon (C), hydrogen (H), an oxygen (O) atoms— though in a dizzying variety of molecular forms. The goal in gasification is to break down this wide variety of forms into the simple fuel gasses of H_2 and CO— hydrogen and carbon monoxide.

Both hydrogen and carbon monoxide are burnable fuel gasses. We do not usually think of carbon monoxide as a fuel gas, but it actually has very good combustion characteristics (despite its poor characteristics when interacting with human hemoglobin). Carbon monoxide and hydrogen have about the same energy density by volume. Both are very clean burning as they only need to take on one oxygen atom, in one simple step, to arrive at the proper end states of combustion, CO_2 and H_2O. This is why an engine run on producer gas can have such clean emissions. The engine becomes the "afterburner" for the more dirty and difficult early stages of combustion that now are handled in the gasifier.

Thus pyrolysis is the application of heat to biomass in the absence of air/oxygen. The volatiles in the biomass are evaporated off as tar gases, and the fixed carbon-to-carbon chains are what remains— otherwise known as charcoal.

Cracking

Cracking is the process of breaking down large complex molecules such as tar into lighter gases by exposure to heat. This process is crucial for the production of clean gas that is compatible with an internal combustion engine because tar gases condense into sticky tar that will rapidly foul the valves of an engine. Cracking is also necessary to ensure proper combustion because complete combustion only occurs when combustible gases thoroughly mix with oxygen. In the course of combustion, the high temperatures produced decompose the large tar molecules that pass through the combustion zone.

Reduction

Reduction is the process of stripping oxygen atoms off combustion products of hydrocarbon (HC) molecules, so as to return the molecules to forms that can burn again. Reduction is the direct reverse process of combustion. Combustion is the combination of combustible gases with oxygen to release heat, producing water vapor and carbon dioxide as waste products. Reduction is the removal of oxygen from these waste products at high temperature to produce combustable gases. Combustion and Reduction are equal and opposite reactions. In fact, in most burning environments, they are both operating simultaneously, in some form of dynamic equilibrium, with repeated movement back and forth between the two processes.

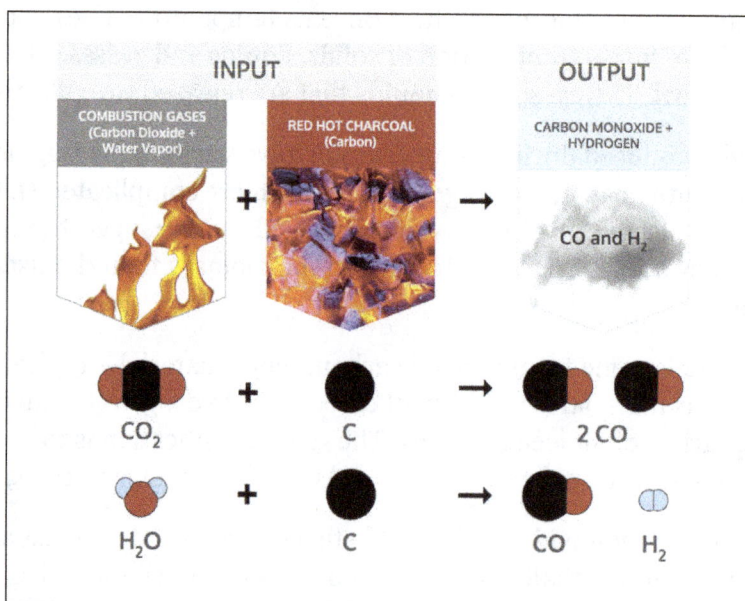

Reduction in a gasifier is accomplished by passing carbon dioxide (CO_2) or water vapor (H_2O) across a bed of red hot charcoal (C). The carbon in the hot charcoal is highly reactive with oxygen; it has such a high oxygen affinity that it strips the oxygen off water vapor and carbon dioxide, and redistributes it to as many single bond sites as possible. The oxygen is more attracted to the bond site on the C than to itself, thus no free oxygen can survive in its usual diatomic O_2 form. All available oxygen will bond to available C sites as individual O until all the oxygen is gone. When all the available oxygen is redistributed as single atoms, reduction stops.

Through this process, CO_2 is reduced by carbon to produce two CO molecules, and H_2O is reduced by carbon to produce H_2 and CO. Both H_2 and CO are combustable fuel gases, and those fuel gasses can then be piped off to do desired work elsewhere.

Combustion and Drying

These are the most easily understood of the Five Processes of Gasification. They do what we think by common understanding, though now they do it in the service of Pyrolysis and Reduction.

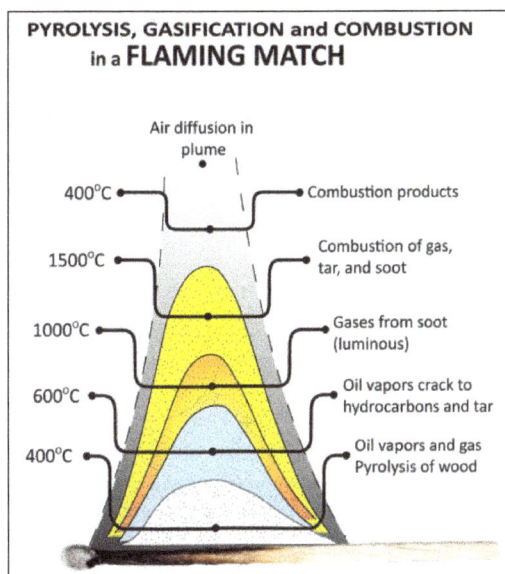

Combustion is the only net exothermic process of the Five Processes of Gasification; ultimately, all of the heat that drives drying, pyrolysis, and reduction comes either directly from combustion, or is recovered indirectly from combustion by heat exchange processes in a gasifier. Combustion can be fueled by either the tar gasses or char from Pyrolysis. Different reactor types use one or the other or both. In a downdraft gasifier, we are trying to burn the tar gasses from pyrolysis to generate heat to run reduction, as well as the CO_2 and H_2O to reduce in reduction. The goal in combustion in a downdraft is to get good mixing and high temps so that all the tars are either burned or cracked, and thus will not be present in the outgoing gas. The char bed and reduction contribute a relatively little to the conversion of messy tars to useful fuel gasses. Solving the tar problem is mostly an issue of tar cracking in the combustion zone.

Drying is what removes the moisture in the biomass before it enters Pyrolysis. All the moisture needs to be (or will be) removed from the fuel before any above 100 °C processes happen. All of the water in the biomass will get vaporized out of the fuel at some point in the higher temp processes. Where and how this happens is one of the major issues that has to be solved for successful gasification. High moisture content fuel, and/or poor handling of the moisture internally, is one of the most common reasons for failure to produce clean gas.

More simply you might just think of gasification as burning a match, but interrupting the process by piping off the clear gas not letting it mix with oxygen and complete combustion. Or you might think of it as running your car engine extremely rich, creating enough heat to break apart the raw fuel, but without enough oxygen to complete combustion, thus sending burnable gasses out the exhaust. This is how a hot rodder gets flames out the exhaust pipes.

Methanol Economy

The methanol economy is a suggested future economy in which methanol and dimethyl ether replace fossil fuels as a means of energy storage, ground transportation fuel, and raw material for synthetic hydrocarbons and their products. It offers an alternative to the proposed hydrogen economy or ethanol economy.

In the 1990s, Nobel prize winner George A. Olah advocated a methanol economy; in 2006, he and two co-authors, G. K. Surya Prakash and Alain Goeppert, published a summary of the state of fossil fuel and alternative energy sources, including their availability and limitations, before suggesting a methanol economy.

Methanol can be produced from a wide variety of sources including still-abundant fossil fuels (natural gas, coal, oil shale, tar sands, etc.) as well as agricultural products and municipal waste, wood and varied biomass. It can also be made from chemical recycling of carbon dioxide.

Uses

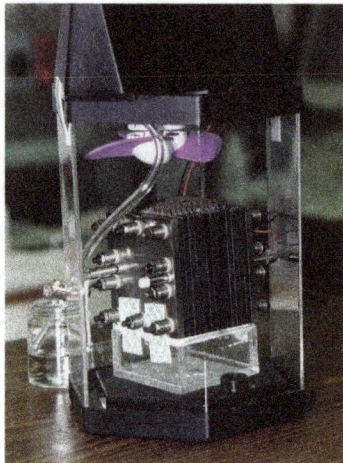

Direct-methanol fuel cell.

Fuel

Methanol is a fuel for heat engines and fuel cells. Due to its high octane rating it can be used directly as a fuel in flex-fuel cars (including hybrid and plug-in hybrid vehicles) using existing internal combustion engines (ICE). Methanol can also be burned in some other kinds of engine or to provide heat as other liquid fuels are used. Fuel cells, can use methanol either directly in Direct Methanol Fuel Cells (DMFC) or indirectly (after conversion into hydrogen by reforming).

Feedstock

Methanol is already used today on a large scale to produce a variety of chemicals and products. Global methanol demand as a chemical feedstock reached around 42 million metric tonnes per year as of 2015. Through the methanol-to-gasoline (MTG) process, it can be transformed into gasoline. Using the methanol-to-olefin (MTO) process, methanol can also be converted to ethylene and propylene, the two chemicals produced in largest amounts by the petrochemical industry. These are important building blocks for the production of essential polymers (LDPE, HDPE, PP) and like other chemical intermediates are currently produced mainly from petroleum feedstock. Their production from methanol could therefore reduce our dependency on petroleum. It would also make it possible to continue producing these chemicals when fossil fuels reserves are depleted.

Production

Today most methanol is produced from methane through syngas. Trinidad and Tobago is currently the world's largest methanol exporter, with exports mainly to the United States. The natural gas that serves as feedstock for the production of methanol comes from the same sources as other uses. Unconventional gas resources such as coalbed methane, tight sand gas and eventually the very large methane hydrate resources present under the continental shelves of the seas and Siberian and Canadian tundra could also be used to provide the necessary gas.

The conventional route to methanol from methane passes through syngas generation by steam reforming combined (or not) with partial oxidation. New and more efficient ways to convert methane into methanol are also being developed. These include:

- Methane oxidation with homogeneous catalysts in sulfuric acid media.
- Methane bromination followed by hydrolysis of the obtained bromomethane.
- Direct oxidation of methane with oxygen.
- Microbial or photochemical conversion of methane.
- Partial methane oxidation with trapping of the partially oxidized product and subsequent extraction on copper and iron exchanged Zeolite (e.g. Alpha-Oxygen).

All these synthetic routes emit the greenhouse gas carbon dioxide CO_2. To mitigate this, methanol can be made through ways minimizing the emission of CO_2. One solution is to produce it from syngas obtained by biomass gasification. For this purpose any biomass can be used including wood, wood wastes, grass, agricultural crops and their by-products, animal waste, aquatic plants and municipal waste. There is no need to use food crops as in the case of ethanol from corn, sugar cane and wheat.

Biomass → Syngas (CO, CO_2, H_2) → CH_3OH

Methanol can be synthesized from carbon and hydrogen from any source, including still available fossil fuels and biomass. CO_2 emitted from fossil fuel burning power plants and other industries and eventually even the CO_2 contained in the air, can be a source of carbon. It can also be made from chemical recycling of carbon dioxide, which Carbon Recycling International has demonstrated with its first commercial scale plant. Initially the major source will be the CO_2 rich flue gases of fossil-fuel-burning power plants or exhaust from cement and other factories. In the longer range however, considering diminishing fossil fuel resources and the effect of their utilization on earth's atmosphere, even the low concentration of atmospheric CO_2 itself could be captured and recycled via methanol, thus supplementing nature's own photosynthetic cycle. Efficient new absorbents to capture atmospheric CO_2 are being developed, mimicking plants' ability. Chemical recycling of CO_2 to new fuels and materials could thus become feasible, making them renewable on the human timescale.

Methanol can also be produced from CO_2 by catalytic hydrogenation of CO_2 with H_2 where the hydrogen has been obtained from water electrolysis. This is the process used by Carbon Recycling International of Iceland. Methanol may also be produced through CO_2 electrochemical reduction, if electrical power is available. The energy needed for these reactions in order to be carbon neutral would come from renewable energy sources such as wind, hydroelectricity and solar as well as nuclear power. In effect, all of them allow free energy to be stored in easily transportable methanol, which is made immediately from hydrogen and carbon dioxide, rather than attempting to store energy in free hydrogen.

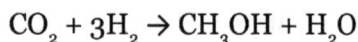

$$CO_2 + 3H_2 \rightarrow CH_3OH + H_2O$$

or with electric energy

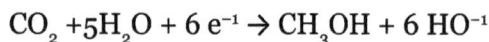

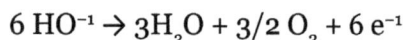

$$CO_2 + 5H_2O + 6 e^{-1} \rightarrow CH_3OH + 6 HO^{-1}$$

$$6 HO^{-1} \rightarrow 3H_2O + 3/2 O_2 + 6 e^{-1}$$

Total:

$$CO_2 + 2H_2O + electric\ energy \rightarrow CH_3OH + 3/2 O_2$$

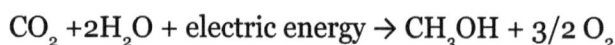

The necessary CO_2 would be captured from fossil fuel burning power plants and other industrial flue gases including cement factories. With diminishing fossil fuel resources and therefore CO_2 emissions, the CO_2 content in the air could also be used. Considering the low concentration of CO_2 in air (0.04%) improved and economically viable technologies to absorb CO_2 will have to be developed. For this reason, extraction of CO_2 from water could be more feasible due to its higher concentrations in dissolved form. This would allow the chemical recycling of CO_2, thus mimicking nature's photosynthesis.

Advantages

In the process of photosynthesis, green plants use the energy of sunlight to split water into free oxygen (which is released) and free hydrogen. Rather than attempt to store the hydrogen, plants immediately capture carbon dioxide from the air to allow the hydrogen to reduce it to storable fuels such as hydrocarbons (plant oils and terpenes) and polyalcohols (glycerol, sugars and starches). In the methanol economy, any process which similarly produces free hydrogen, proposes to immediately use it "captively" to reduce carbon dioxide into methanol, which, like plant products from photosynthesis, has great advantages in storage and transport over free hydrogen itself.

Methanol is a liquid under normal conditions, allowing it to be stored, transported and dispensed easily, much like gasoline and diesel fuel. It can also be readily transformed by dehydration into dimethyl ether, a diesel fuel substitute with a cetane number of 55.

Methanol is water-soluble: An accidental release of methanol in the environment would, cause much less damage than a comparable gasoline or crude oil spill. Unlike these fuels, methanol is biodegradable and totally soluble in water, and would be rapidly diluted to a concentration low enough for microorganism to start biodegradation. This effect is already exploited in water treatment plants, where methanol is already used for denitrification and as a nutrient for bacteria. Accidental release causing groundwater pollution has not been thoroughly studied yet, though it is believed that it might undergo relatively rapid.

Comparison with Hydrogen

Methanol economy advantages compared to a hydrogen economy:

- Efficient energy storage by volume, as compared with compressed hydrogen. When hydrogen pressure-confinement vessel is taken into account, an advantage in energy storage by weight can also be realized. The volumetric energy density of methanol is considerably higher than liquid hydrogen, in part because of the low density of liquid hydrogen of 71 grams/litre. Hence there is actually more hydrogen in a litre of methanol (99 grams/litre) than in a litre of liquid hydrogen, and methanol needs no cryogenic container maintained at a temperature of -253 °C .

- A liquid hydrogen infrastructure would be prohibitively expensive. Methanol can use existing gasoline infrastructure with only limited modifications.

- Can be blended with gasoline (for example in M85, a mixture containing 85% methanol and 15% gasoline).

- User friendly: Hydrogen is volatile, and its confinements uses high pressure or cryogenic systems.

- Less losses: Hydrogen leaks more easily than methanol. Heat will evaporate liquid hydrogen, giving expected losses up to 0.3% per day in storage tanks.

Comparison with Ethanol

- Can be made from any organic material using proven technology going through syngas. There is no need to use food crops and compete with food production. The amount of methanol that can be generated from biomass is much greater than ethanol.

- Can compete with and complement ethanol in a diversified energy marketplace. Methanol obtained from fossil fuels has a lower price than ethanol.

- Can be blended in gasoline like ethanol. In 2007, China blended more than 1 billion US gallons (3,800,000 m³) of methanol into fuel and will introduce methanol fuel standard by mid-2008. M85, a mixture of 85% methanol and 15% gasoline can be used much like E85 sold in some gas stations today.

Disadvantages

- High energy costs currently associated with generating and transporting hydrogen offsite.

- Presently generated from natural gas still dependent on fossil fuels (although any combustible hydrocarbon can be used).

- Energy density (by weight or volume) one half of that of gasoline and 24% less than ethanol

- Handling

  ◦ If no inhibitors are used, methanol is corrosive to some common metals including aluminum, zinc and manganese. Parts of the engine fuel-intake systems are made from aluminum. Similar to ethanol, compatible material for fuel tanks, gasket and engine intake have to be used.

  ◦ As with similarly corrosive and hydrophilic ethanol, existing pipelines designed for petroleum products cannot handle methanol. Thus methanol requires shipment at higher energy cost in trucks and trains, until new pipeline infrastructure can be built, or existing pipelines are retrofitted for methanol transport.

    ◦ Methanol, as an alcohol, increases the permeability of some plastics to fuel vapors (e.g. high-density polyethylene). This property of methanol has the possibility of increasing emissions of volatile organic compounds (VOCs) from fuel, which contributes to increased tropospheric ozone and possibly human exposure.

- Low volatility in cold weather: Pure methanol-fueled engines can be difficult to start, and they run inefficiently until warmed up. This is why a mixture containing 85% methanol and 15% gasoline called M85 is generally used in ICEs. The gasoline allows the engine to start even at lower temperatures.

- With the exception of low level exposure, methanol is toxic: Methanol is lethal when ingested in larger amounts (30 to 100 mL). But so are most motor fuels, including gasoline (120 to 300 mL) and diesel fuel. Gasoline also contains small amounts of many compounds known to be carcinogenic (e.g. benzene). Methanol is not a carcinogen, nor does it contain carcinogens. However, methanol may be metabolized in the body to formaldehyde, which is both toxic and carcinogenic. Methanol occurs naturally in small quantities in the human body and in edible fruits.

- Methanol is a liquid: This creates a greater fire risk compared to hydrogen in open spaces as Methanol leaks do not dissipate. Methanol burns invisibly unlike gasoline. Compared to gasoline, however, methanol is much safer. It is more difficult to ignite and releases less heat when it burns. Methanol fires can be extinguished with plain water, whereas gasoline floats on water and continues to burn. The EPA has estimated that switching fuels from gasoline to methanol would reduce the incidence of fuel related fires by 90%.

Ethanol Fuel Energy Balance

In order to create ethanol, all biomass needs to go through some of these steps: it needs to be grown, collected, dried, fermented, and burned. All of these steps require resources and an infrastructure. The ratio of the energy released by burning the resulting ethanol fuel to the energy used in the process, is known as the ethanol fuel energy balance (sometimes called "Net energy gain") and studied as part of the wider field of energy economics. Figures compiled in a 2007 National Geographic Magazine article point to modest results for corn ethanol produced in the US: 1 unit of energy input equals 1.3 energy units of corn ethanol energy. The energy balance for sugarcane ethanol produced in Brazil is much more favorable, 1 to 8. Over the years, however, many reports have been produced with contradicting energy balance estimates. A 2006 University of California Berkeley study, after analyzing six separate studies, concluded that producing ethanol from corn uses marginally less petroleum than producing gasoline.

Energy Balance Reports

In 1995 the USDA released a report stating that the net energy balance of corn ethanol in the United States was an average of 1.24. It was previously considered to have a negative net energy balance. However, due to increases in corn crop yield and more efficient farming practices corn ethanol had gained energy efficiency.

Ken Cassman, a professor of agronomy at the University of Nebraska–Lincoln, said in 2008 that ethanol has a substantial net positive direct energy balance—1.5 to 1.6 more units of energy are derived from ethanol than are used to produce it. Comparing 2008 to 2003, Alan Tiemann of Seward, a Nebraska Corn Board member, said that ethanol plants produce 15 percent more ethanol from a bushel of corn and use about 20 percent less energy in the process. At the same time, corn growers are more efficient, producing more corn per acre and using less energy to do so.

Opponents of corn ethanol production in the U.S. often quote the 2005 paper of David Pimentel, a retired Entomologist, and Tadeusz Patzek, a Geological Engineer from UC Berkeley. Both have been exceptionally critical of ethanol and other biofuels. Their studies contend that ethanol, and biofuels in general, are "energy negative", meaning they take more energy to produce than is contained in the final product.

A 2006 article in Science offers the consensus opinion that current corn ethanol technologies had similar greenhouse gas emissions to gasoline, but was much less petroleum-intensive than gasoline. Fossil fuels also require significant energy inputs which have seldom been accounted for in the past.

Ethanol is not the only product created during production. By-products also have energy content. Corn is typically 66% starch and the remaining 33% is not fermented. This unfermented component is called distillers grain, which is high in fats and proteins, and makes a good animal feed supplement.

In 2000, Dr. Michael Wang, of Argonne National Laboratory, wrote that these ethanol by-products are the most contentious issue in evaluating the energy balance of ethanol. He wrote that Pimentel assumes that corn ethanol entirely replaces gasoline and so the quantity of by-products is too large for the market to absorb, and they become waste. At lower quantities of production, Wang finds it appropriate to credit corn ethanol based on the input energy requirement of the feed product or good that the ethanol by-product displaces. In 2004, a USDA report found that co-products accounting made the difference between energy ratios of 1.06 and 1.67. In 2006, MIT researcher Tiffany Groode came to similar conclusions about the co-product issue.

In Brazil where sugar cane is used, the yield is higher, and conversion to ethanol is more energy efficient than corn. Recent developments with cellulosic ethanol production may improve yields even further.

In 2006 a study from the University of Minnesota found that corn-grain ethanol produced 1.25 units of energy per unit put in.

A 2008 study by the University of Nebraska found a 5.4 energy balance for ethanol derived specifically from switchgrass. This estimate is better than in previous studies and according to the authors partly due to the larger size of the field trial (3-9 ha) on 10 farms.

Variables

According to DoE, to evaluate the net energy of ethanol four variables must be considered:

1.   The amount of energy contained in the final ethanol product.

2. The amount of energy directly consumed to make the ethanol (such as the diesel used in tractors).

3. The quality of the resulting ethanol compared to the quality of refined gasoline.

4. The energy indirectly consumed (in order to make the ethanol processing plant, etc.).

Much of the current academic discussion regarding ethanol currently revolves around issues of system borders. This refers to how complete of a picture is drawn for energy inputs. There is debate on whether to include items like the energy required to feed the people tending and processing the corn, to erect and repair farm fences, even the amount of energy a tractor represents.

In addition, there is no consensus on what sort of value to give the rest of the corn (such as the stalk), commonly known as the 'coproduct.' Some studies leave it on the field to protect the soil from erosion and to add organic matter, while others take and burn the coproduct to power the ethanol plant, but do not address the resulting soil erosion (which would require energy in the form of fertilizer to replace). Depending on the ethanol study you read, net energy returns vary from .7-1.5 units of ethanol per unit of fossil fuel energy consumed. For comparison, that same one unit of fossil fuel invested in oil and gas extraction (in the lower 48 States) will yield 15 units of gasoline, a yield an order of magnitude better than current ethanol production technologies, ignoring the energy quality arguments above and the fact that the gain (14 units) is both declining and not carbon neutral.

In this regard, geography is the decisive factor. In tropical regions with abundant water and land resources, such as Brazil and Colombia, the viability of production of ethanol from sugarcane is no longer in question; in fact, the burning of sugarcane residues (bagasse) generates far more energy than needed to operate the ethanol plants, and many of them are now selling electric energy to the utilities. However, while there may be a positive net energy return at the moment, recent research suggests that the sugarcane plantations are not sustainable in the long run, as they are depleting the soil of nutrients and carbon matterOn the other hand, productivity of sugar cane per land area in Brazil has consistently grown over the decades; sugar cane has been shown to be less depleting to the soil than cattle and yearly cultures; and there are many regions in the country where sugar cane has been cultivated for centuries. Those facts suggest that related soil depletion processes are very slow and therefore ethanol from sugar cane may be far more sustainable in the long run than common fossil fuel alternatives. Besides, since the energy surplus is high in the case of sugar cane ethanol, conceivably part of that energy can be used to synthesize fertilizers and replenish soil depletion a long time, therefore making the process indefinitely sustainable.

The picture is different for other regions, such as most of the United States, where the climate is too cool for sugarcane. In the U.S., agricultural ethanol is generally obtained from grain, chiefly corn. But it can also be obtained from cellulose, a more energy balanced bioethanol.

Clean Production Bioethanol

Clean production bioethanol is a biofuel obtained by maximizing non-greenhouse gas emitting (renewable) resources:

- Energy directly consumed to make the ethanol is renewable energy. The farm equipment and ethanol plant use an ethanol engine, biodiesel, air engine or electricity cogenerated during ethanol production, or even wind power and solar energy.

- Energy indirectly consumed is, as much as possible, renewable. Examples would be reducing either the amount or fossil carbon content of applied pest control chemicals and fertilizers, or accomplishing deliveries of farm inputs or of finished bioethanol fuel to market that minimize the use of fossil fuels. Optimally located biomass and ethanol production must balance many factors: minimizing distances to and from markets, effectively collecting and employing biomass wastes, maximizing crop yields based on enduring soil quality, available natural pest control and adequate sun and water, and optimizing a sufficient mix and rotation of plant species on cultivated, fallow and preserved land for human, animal and energy consumption.

Using ethanol returns carbon to the atmosphere whereas burning gasoline adds carbon to the atmosphere. Thus the effects of gasoline burning increase over time.

References

- Biorefinery: bioenergyconsult.com, Retrieved 13 July, 2019

- Venkata mohan, s., lalit babu, v., srikanth, s., sarma, p.n., 2008d. "bio-electrochemical behavior of fermentative hydrogen production process with the function of feeding ph". International journal of hydrogen energy doi:10.1016/j.ijhydene.2008.05.073

- Gasification: studentenergy.org, Retrieved 17 February, 2019

- Müller, volker (2001). "bacterial fermentation" (pdf). Els. John wiley & sons, ltd. Doi:10.1038/npg.els.0001415. Isbn 9780470015902

- Dubé, marc a, et al. (2007). "acid-catalyzed transesterification of canola oil to biodiesel under single- and two-phase reaction conditions". Energy & fuels 21: 2450–2459. American chemical society. Retrieved on 2007-11-01

- Uyar b (september 2016). "bioreactor design for photofermentative hydrogen production". Bioprocess and biosystems engineering. 39 (9): 1331–40. Doi:10.1007/s00449-016-1614-9. Pmid 27142376

- "Green car congress: conocophillips begins production of renewable diesel fuel at whitegate refinery". Greencarcongress.com. 2012. Retrieved december 27, 2012

- Gasification-explained: allpowerlabs.com, Retrieved 24 March, 2019

- M. R. Schmer, k. P. Vogel, r. B. Mitchell, and r. K. Perrin net energy of cellulosic ethanol from switchgrass pnas published january 7, 2008, doi:10.1073/pnas.0704767105

PERMISSIONS

All chapters in this book are published with permission under the Creative Commons Attribution Share Alike License or equivalent. Every chapter published in this book has been scrutinized by our experts. Their significance has been extensively debated. The topics covered herein carry significant information for a comprehensive understanding. They may even be implemented as practical applications or may be referred to as a beginning point for further studies.

We would like to thank the editorial team for lending their expertise to make the book truly unique. They have played a crucial role in the development of this book. Without their invaluable contributions this book wouldn't have been possible. They have made vital efforts to compile up to date information on the varied aspects of this subject to make this book a valuable addition to the collection of many professionals and students.

This book was conceptualized with the vision of imparting up-to-date and integrated information in this field. To ensure the same, a matchless editorial board was set up. Every individual on the board went through rigorous rounds of assessment to prove their worth. After which they invested a large part of their time researching and compiling the most relevant data for our readers.

The editorial board has been involved in producing this book since its inception. They have spent rigorous hours researching and exploring the diverse topics which have resulted in the successful publishing of this book. They have passed on their knowledge of decades through this book. To expedite this challenging task, the publisher supported the team at every step. A small team of assistant editors was also appointed to further simplify the editing procedure and attain best results for the readers.

Apart from the editorial board, the designing team has also invested a significant amount of their time in understanding the subject and creating the most relevant covers. They scrutinized every image to scout for the most suitable representation of the subject and create an appropriate cover for the book.

The publishing team has been an ardent support to the editorial, designing and production team. Their endless efforts to recruit the best for this project, has resulted in the accomplishment of this book. They are a veteran in the field of academics and their pool of knowledge is as vast as their experience in printing. Their expertise and guidance has proved useful at every step. Their uncompromising quality standards have made this book an exceptional effort. Their encouragement from time to time has been an inspiration for everyone.

The publisher and the editorial board hope that this book will prove to be a valuable piece of knowledge for students, practitioners and scholars across the globe.

INDEX